1st World Energy Materials Conference 2023

Journal of Physics: Conference Series Volume 2749

Shenzhen, China
25 – 27 November 2023

ISBN: 978-1-7138-9736-1
ISSN: 1742-6588

Printed from e-media with permission by:

Curran Associates, Inc.
57 Morehouse Lane
Red Hook, NY 12571

Some format issues inherent in the e-media version may also appear in this print version.

Institute of Physics
Dirac House, Temple Back
Bristol BS1 6BE UK

Phone: 44 1 17 929 7481
Fax: 44 1 17 920 0979

techtracking@iop.org

Additional copies of this publication are available from:

Curran Associates, Inc.
57 Morehouse Lane
Red Hook, NY 12571 USA
Phone: 845-758-0400
Fax: 845-758-2633
Email: curran@proceedings.com
Web: www.proceedings.com

TABLE OF CONTENTS

Preface

This is the proceedings of the "1st World Energy Materials Conference", which was held in Shenzhen, China on November 25-27, 2023. There are 24 forums in the conference. This collection consists of papers in new energy materials, new material industry development trends and hot spots, green low-carbon energy technology, development and challenges of renewable energy, etc., to discuss the latest progress and innovation in the field of energy materials, with a view to contributing wisdom to the promotion of world energy development, material science and technology innovation exchanges and cooperation, as well as high-quality and sustainable development.

The list of Committees:

Honorary President: Lian Zhou (Academician of Chinese Academy of Engineering); Boyun Huang (Academician of Chinese Academy of Engineering); Paul Siffert (General Secretary of European Materials Research Society)

President: Yuanyuan Li (President of Chinese Materials Research Society); Bingbo Wei (President of Chinese Materials Research Society); Anthony Kenyon (President of European Materials Research Society)

Vice President: Liquan Chen (Academician of Chinese Academy of Engineering); Jianxin Xie (Academician of Chinese Academy of Engineering); Feng Wu (Beijing Institute of Technology); Fusheng Pan (Chongqing University); Wenjiang Ding (Shanghai Jiaotong University); Zuoren Nie (Beijing University of Technology); Pingxiang Zhang (Northwest Nonferrous Metals Research Institute); Zhigang Zou (Nanjing University); Daqiao Meng (China Academy of Engineering Physics); Zhonglin Wang (Beijing Institute of Nanoenergy and Nanosystem, CAS); Cewen Nan (Tsinghua University); Jin Zhang (Beijing University); Huiming Cheng (Shenzhen Institute of Advanced Technology, CAS); Xiaoping Ouyang (Northwest Institute of Nuclear Technology); Min Zhu (South China University of Technology); Henry H. Radamson (Institute of Microelectronics of the CAS); Joan Ramon Morante (Catalonia Institute for Energy Research IREC); Jean-Marie Tarascon (Le Collège de France); Patrice Simon (Paul Sabatier University); Franck Tessier (CNRS Research Director); Madoka TAKAI (The University of Tokyo)

Peer Review Statement

All papers published in this volume have been reviewed through processes administered by the Editors. Reviews were conducted by expert referees to the professional and scientific standards expected of a proceedings journal published by IOP Publishing.

- **Type of peer review:** Single Anonymous
- **Conference submission management system:** Morressier
- **Number of submissions received:** 14
- **Number of submissions sent for review:** 14
- **Number of submissions accepted:** 14
- **Acceptance Rate (Submissions Accepted / Submissions Received × 100):** 100
- **Average number of reviews per paper:** 2
- **Total number of reviewers involved:** 10
- **Contact person for queries:**
  Name: Yan Hao
  Email: hy_cmrs@163.com
  Affiliation: Chinese Materials Research Society

Journal of Physics: Conference Series **2749** (2024) 012001 doi:10.1088/1742-6596/2749/1/012001

Phase-field simulation of the effect of grain boundary on fission gas migration in UO_2 fuel

Dongkun Liu, Qingyu Wang*, Tian Zhang

Harbin Engineering University, Harbin, Heilongjiang, 150001, China

wangqingyu@hrbeu.edu.cn

Abstract. Grain boundaries are widely recognized as line defect sinks. During reactor operation, vacancies and fission gas atoms within UO_2 fuel grains migrate to the grain boundaries, forming bubbles at these locations. In order to better understand the effect of grain boundaries on the migration of fission gas atoms, this study employed a phase-field model to simulate the nucleation and growth process of fission gas bubbles within the UO_2 fuel grains and in the vicinity of grain boundaries. This study also investigated the degree to which grain boundaries affect fission gas atoms under different temperature conditions. The results indicated that in models containing grain boundaries, the nucleation of fission gas bubbles occurred earlier, as compared to models without grain boundaries. A noticeable bubble denuded zone was also observed adjacent to grain boundary interfaces. Furthermore, with increasing temperature, the bubble denuded zone becomes thicker. The effects of irradiation, vacancy diffusion, Ostwald ripening, as well as grain boundary trapping were discussed.

1. Introduction

During the operation of the reactor, various fission products such as inert gases Xe and Kr, are continuously produced as fission progresses. These two gases produce approximately 0.25-0.3 atoms per fission [1], and the production rate of Xe is nearly 10 times that of Kr [2].Due to the low solubility of these two gases in UO_2, most of Xe and Kr diffuse to form bubbles within or between crystals [3],[4]. As bubbles increase and grow, intracrystalline bubbles can cause swelling of fuel pellets [3], and intergranular bubbles connect to form channels, causing fission gas to diffuse into free space along the channels, resulting in the release of fission gas [4].

The generation and release of fission gas is one of the key constraints for improving the burnup level of reactors. The accumulation of fission gas in reactor fuel rods, especially when the reactor burnup reaches a certain level, can cause an increase in internal pressure and stress in the reactor fuel rods,ultimately resulting in deformation of the fuel rod cladding and the rupture of the cladding, causing the release of radioactive materials [5]. Therefore, studying the release laws of fission gases can assist in improving fuel design, enhancing reactor safety, and increasing overall reactor economics [6].

Studies on the release behavior of fission gases often employ the rate theory model, which uses the spatial mean-field method to improve computational efficiency but ignores the evolution of various microstructure features. Compared to the rate theory method, the phase-field method can not only investigate thermodynamic and dynamic factors, but also track the dynamic evolution process of the model. Based on Ginzburg-Landau theory, it establishes a mathematical model that can accurately describe the evolution of the system over time through partial differential equations [7], to obtain the instantaneous state of the system in time and space. The phase field method avoids the tracking problem of complex interfaces by incorporating different interface descriptions into its formulas, and is easily coupled with other physical fields (such as noise field, velocity field, composition field, etc.) [8].

There have been many related studies using the phase-field method to simulate the release behavior of fission gases in UO_2 fuel. Hu et al. [9] constructed a phase field model containing vacancy concentration field and interstitial atom concentration field using the Cahn Hilliard equation. The model simulated the evolution of voids under irradiation conditions and obtained a series of parameters such as critical radius and quantity density of voids. Millett et al. [10] constructed a phase field model including the Cahn-Hilliard equation and the Allen-Cahn equation. This equation system distinguishes different phase boundaries by controlling order parameters, allowing for different dynamic and thermodynamic descriptions of different phases. Through this model, three stages of void development were obtained. In addition, the model also demonstrated that grain boundaries act as absorption sinks for defects and have strong absorption effects on vacancies and interstitial atoms. Void denuded and peak zones were observed near the grain boundaries. However, the parameters selected in this model are not for specific materials, and fission gas atoms were not taken into account when considering the role of grain boundaries as absorption sinks, which cannot quantitatively analyze the release law of fission gases near UO_2 fuel grain boundaries.

This study also used a phase-field model and referred to the above research. The nucleation and growth process of fission gas bubbles inside and near UO_2 fuel grains under irradiation conditions were simulated using a parabolic approximation free energy functional, specific diffusion coefficient values, and micron scale grain size. The influence of grain boundaries on the release of fission gas in UO_2 fuel was studied, and the degree of influence of grain boundaries on different temperatures was investigated.

2. Phase-field model

2.1. Definition of independent variables

This model investigates the concentration distribution of point defect vacancies and fission gas generated by irradiation effects in polycrystalline materials. Two conserved field variables are introduced: $c_v(\boldsymbol{r}, t)$, $c_g(\boldsymbol{r}, t)$, which represent the vacancy mole fraction and fission gas mole fraction. This mole fraction c can be determined by $c = \rho V_a$, where ρ is the number density of vacancies or fission gas atoms, V_a is the volume of U atoms in UO_2 crystal cells, with a value of 0.0409 nm ^{3} [11].

As mentioned in the introduction, fission gases in fuel matrix forms bubbles through diffusion, and there are two stable phases: the solid fuel matrix phase and the fission gas bubble phase, using the order parameter η distinguish between the two phases: when $\eta = 0$ is the matrix phase, $\eta = 1$ is the bubble phase. There is an equilibrium concentration of vacancies and gases in the matrix phase and bubble phase. For the matrix phase, the equilibrium concentration of vacancies and gases is represented by the following equation:

$$c_v^{m,eq} = \exp\left(-\frac{E_v^f}{k_B T}\right) \tag{1}$$

$$c_g^{m,eq} = \exp\left(-\frac{E_g^f}{k_B T}\right) \tag{2}$$

where E_v^f and E_g^f are the formation energies of vacancies and fission gas atomic, $E_v^f = 5.1$eV[12]. The dissolution energy of Xe atoms in vacancies E_g^s =5.21eV[13], choose $E_g^f = E_v^f + E_g^s = 10.31$eV. k_B is the Boltzmann constant, and T is the temperature. We take the gas equilibrium concentration $c_g^{b,eq}$ = 0.494[11], when $c_g = 0.494$, the van der Waals free energy of the bubble phase is the lowest. Due to the fluctuation of gas concentration cg during the simulation process, in order to maintain $c_v + c_g \leq 1$ in the bubble phase, the vacancy equilibrium concentration is taken as $c_v^{b,eq} = 0.9 - c_g$, $c_v^{b,eq}$ will decreases as c_g increases.

This phase-field model focuses on the release behavior of fission gases in polycrystalline fuels, using order parameters ϕ_i (i=1, 2,..., P) distinguishes grains with different orientations. Within the grains with the i-th orientation ϕ_i=1, in other oriented grain phases ϕ_i=0. Near the contact surface of grains with different orientations, ϕ_i is approximately equal to 0.5. Through investigating $\triangleleft = \sum_{i=1}^{P} \phi_i^2$, it is possible to distinguish between the interior and boundaries of the grain, when $\triangleleft = 1$ is within the grain, $\triangleleft \approx 0.5$ is on the grain boundary.

2.2. Free energy functional

The phase-field method uses the total free energy functional F composed of the free energies of each phase and interface, which is consistent with the free energy of non-uniform systems defined by Cahn Hilliard[14]. The free energy functional F form used in this study is as follows:

$$F = \int_V \begin{matrix} h(\eta) f^{solid}(c_v, c_g) + j(\eta) f^{bubble}(c_v, c_g) \\ + f^{poly}(\eta, \phi_{1\to P}) + f^{grad}(c_v, c_g, \eta, \phi_{1\to P}) \end{matrix} dV \tag{3}$$

Where f^{solid} is the matrix phase free energy density function, f^{bubble} is the bubble phase free energy density function, f^{poly} is the polycrystalline free energy density function, and f^{grad} is the gradient contribution term. $h(\eta) = (1-\eta)^2$、$j(\eta) = \eta^2$ is the weighting function. When $\eta = 0$, the contribution of f^{olid} to the total free energy reaches its maximum, and f^{bubble} has no contribution. When $\eta = 1$ the opposite is true.

To simplify the numerical calculation of the control equation, parabolic approximations were applied to the matrix phase free energy density function and the bubble phase free energy density function, which are similar to the free energy density function used by Li et al.[15]. And make a simple correction to ensure that the free energy at the grain boundary is less than the free energy within the grain. The approximate expression of the matrix phase free energy density function f^{solid} is as follows:

$$\begin{aligned} f^{solid}(c_v, c_g) &= f^{solid,v}(c_v) + f^{solid,g}(c_g) \\ &= \frac{1}{2} f^{m,fix} k_v^m (c_v - c_v^{m,eq})^2 + \frac{1}{2} f^{m,fix} k_g^m (c_g - c_g^{m,eq})^2 \end{aligned} \tag{4}$$

The correction term is $f^{m,fix} = \frac{4}{5} + \frac{1}{5}\triangleleft$, assuming that the free energy density of vacancies and gases at the grain boundary of the matrix phase is about nine tenths of that within the grain, this leads to a trend of diffusion of vacancies and fission gases towards the grain boundary.

The curvature of the parabola is controlled by k_v^m and k_g^m in the expression of the free energy density function, as follows:

$$k_v^m = \frac{1}{(c_v^0 - c_v^{m,eq})} \left[\frac{RT}{V_m} \left[\ln c_v^0 - \ln(1 - c_v^0) \right] + \frac{N_A E_v^f}{V_m} \right] \tag{5}$$

$$k_g^m = \frac{1}{(c_g^0 - c_g^{m,eq})} \left[\frac{RT}{V_m} \left[\ln c_g^0 - \ln(1 - c_g^0) \right] + \frac{N_A E_g^f}{V_m} \right] \tag{6}$$

Assuming $c_v^0 = c_g^0 = 0.01$[15]. $k_v^b = 1.7999 \times 10^{12}$ J/m^3, $k_g^b = 3.8408 \times 10^{12}$ J/m^3, calculated at T=1276K.

Similarly, the approximate expression for the free energy density function f^{bubble} of the bubble phase is as follows:

$$\begin{aligned} f^{bubble}(c_v, c_g) &= f^{bubble,v}(c_v) + f^{bubble,g}(c_g) \\ &= \frac{1}{2} f_v^{b,fix} k_v^b (c_v - c_v^{b,eq})^2 + \frac{1}{2} f_g^{b,fix} k_g^b (c_g - c_g^{b,eq})^2 \end{aligned} \tag{7}$$

Correction term:

$$f_v^{b,fix} = \begin{cases} 1 & c_v \le c_v^{b,eq} \\ \left(1 + c_v - c_v^{b,eq}\right)^{10} & c_v > c_v^{b,eq} \end{cases} \tag{8}$$

$$f_g^{b,fix} = \begin{cases} 1 & c_g \le c_g^{b,eq} \\ \left(1 + c_g - c_g^{b,eq}\right)^{10} & c_g > c_g^{b,eq} \end{cases} \tag{9}$$

This correction term is added based on the characteristic of the Helmholtz free energy of the bubble phase as the gas concentration increases. When the gas concentration c_g approaches the equilibrium concentration of 0.494, the Helmholtz free energy will sharply increase [11]. Assuming that the vacancy concentration c_g approaches 1, the free energy also exhibits similar characteristics. Take the curvature of f^{solid}, k_g^b = 2.25 × 10^9J/m^3. The variation of the free energy controlled by this curvature with concentration matches well with the Helmholtz free energy in the range of $0<c_g<0.494$. Another curvature $k_v^b = 2 \times 10^{10}$ J/m^3, this is assuming that the vacancy concentration c_v in the matrix phase does not exceed the value set by 0.02 when bubbles nucleate. The variation of matrix phase free energy density and bubble phase free energy density with concentration is shown in figure 1.

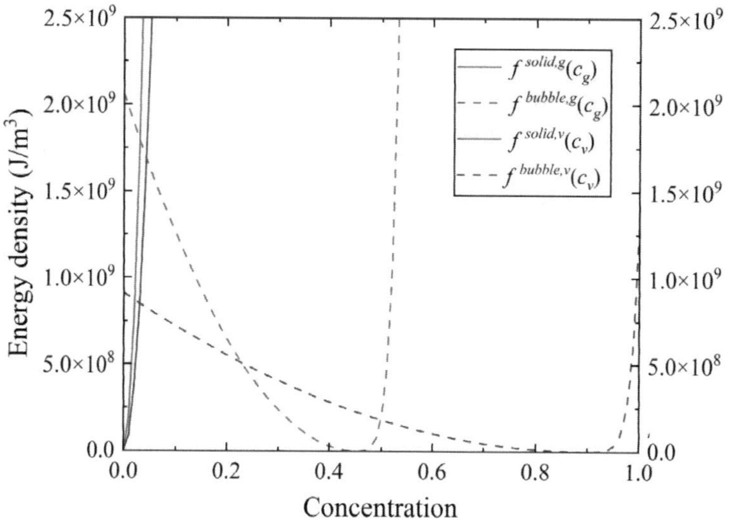

Figure 1. The variation of free energy density with concentration (c_v, c_g) distribution, the abscissa of the lowest point of the curve is the equilibrium concentration. As c_g increases, $c_v^{b,eq}$ gradually decreases, and the image of $f^{bubble,v}$ (blue dashed line) will shift to the left.

The expression for the free energy density function of polycrystals f^{poly} is as follows:

$$f^{poly}(\eta, \phi_{1\to P}) = m\left[\sum_{i=1}^{P}\left(\frac{\phi_i^4}{4} - \frac{\phi_i^2}{2}\right) + \left(\frac{\eta^4}{4} - \frac{\eta^2}{2}\right) + a_{GB}\sum_{i=1}^{P-1}\sum_{j>i}^{P}\phi_i^2\phi_j^2 + a_s\sum_{i=1}^{P}\phi_i^2\eta^2\right] \tag{10}$$

Where m is the free energy barrier coefficient, a_{GB} and a_s are the diffusion interface coefficients, a_{GB} = 1.2, a_s = 0.8.
The expression for the gradient contribution term f^{poly} is as follows:

$$f^{poly}(\eta,\phi_{1\to P}) = m\left[\sum_{i=1}^{P}\left(\frac{\phi_i^4}{4}-\frac{\phi_i^2}{2}\right)+\left(\frac{\eta^4}{4}-\frac{\eta^2}{2}\right)+a_{GB}\sum_{i=1}^{P-1}\sum_{j>i}^{P}\phi_i^2\phi_j^2+a_s\sum_{i=1}^{P}\phi_i^2\eta^2\right] \tag{11}$$

κ_v, κ_g, κ_η, κ_ϕ are the gradient coefficients of vacancy, gas, and two order parameters. $m = 3.0\times10^7$ J/m^3, $\kappa_\phi = 3.38\times10^{-7}$ J/m^3, $\kappa_v = \kappa_g = \kappa_\eta = 1.69\times10^{-6}$ J/m^3

2.3. Control equation

The control equation of the phase-field model adopts the Cahn-Hilliard equation [16],[17] to control the conserved field variables c_v and c_g, and the Allen-Cahn equation [18],[19] to control the order parameters η, ϕ. The control equations for c_v and c_g are given by:

$$\frac{\partial c_v}{\partial t} = \nabla\cdot\left(M_v\nabla\frac{\delta F}{\delta c_v}\right)+P_v(\boldsymbol{r},t)+\xi_v(\boldsymbol{r},t) \tag{12}$$

$$\frac{\partial c_g}{\partial t} = \nabla\cdot\left(M_g\nabla\frac{\delta F}{\delta c_g}\right)+P_g(\boldsymbol{r},t)+\xi_g(\boldsymbol{r},t) \tag{13}$$

Where M_v and M_g are the mobility of vacancies and gases, and their specific form are given by:

$$M_v = \frac{V_a D_v c_v}{k_B T} \tag{14}$$

$$M_g = \frac{V_a D_g c_g}{k_B T} \tag{15}$$

Where D_v and D_g are the diffusion coefficients of vacancies and gases, assuming $D_v = D_g$. The fission gas diffusion coefficient is given by Turnbull's model [20][21]:

$$D_1 = 7.6\times10^{-10}\cdot\exp\left(-4.68\times10^{-19}(\mathrm{J})/(k_B T)\right) \tag{16}$$

$$D_2 = 5.64\times10^{-25}\sqrt{\dot{F}}\cdot\exp\left(-1.91\times10^{-19}(\mathrm{J})/(k_B T)\right) \tag{17}$$

$$D_3 = 8\times10^{-40}\dot{F} \tag{18}$$

Corresponding to the diffusion of gases under three temperature conditions [29], D_1 is the intrinsic diffusion under high temperature (T > 1650K), D_2 is the radiation enhanced diffusion under intermediate temperature (1650K > T > 1350K), and D_3 is the non-thermal radiation driven diffusion under low temperature (T < 1350K). In our simulation, different diffusion coefficients are used based on different temperatures.
The control equation contains two source terms, P_v and P_g respectively represent the rate of vacancy and gas generation. The rate of fission gas P_g generation is calculated by $P_g = V_a\dot{F}Y$, $\dot{F}$ is the fission rate density and taken as 1.09×10^{13} fissions/(cm^3 s)[11] based on the operating data of a typical pressurized water reactor. Y is the yield of fission gas produced by each fission, taken as 0.27. Calculated $P_g = 1.2037\times10^{10}$ s^{-1}, assuming $P_v = 20P_g = 2.4074\times10^{11}$ s^{-1}. ξ_v and ξ_g are simple random functions that represent the thermal fluctuations of vacancies and gases, in order to achieve non-uniform distribution of concentration. The specific form of ξ_v is $\xi_v = c_v R_v(\boldsymbol{r},t)$, R_v is a random function, ranging from -0.01 to 0.01, ξ_g is similar to ξ_v.

The control equations for order parameter η and ϕ are given by:

$$\frac{\partial \eta}{\partial t} = -L_\eta \frac{\delta F}{\delta \eta} \quad (19)$$

$$\frac{\partial \phi_i}{\partial t} = -L_\phi \frac{\delta F}{\delta \phi_i} \quad (i = 1 \rightarrow P) \quad (20)$$

Where L_η and L_ϕ is the mobility coefficient, $L_\eta = L_\phi = 1.56 \times 10^{11}$ m^3/(J s)[11].
Scaling the model without dimensionless processing, time transformation coefficient $t^* = 1 \times 10^{-5}$, τ is the simulation time, corresponding to the actual time $t = \tau / t^*$, free energy density transformation coefficient $f^* = 0.1$. The diffusion coefficient used in the simulation is $\bar{D} = D / t^*$, mobility is $\bar{L} = L / t^*$, and free energy density is $\bar{f} = f \times f^*$.

3. Results and Analysis

3.1. Effect of grain boundaries on nucleation
This section discusses the effect of grain boundaries on the nucleation of fission gas bubbles under T = 1276K, $0 < \tau < 550$s. Construct two different phase-field models, one without grain boundaries (GB) and the other with a straight GB passing through the center. The conclusion is drawn by calculating the nucleation time, radius, and porosity of the simulated area of the bubble. The simulation area of the model is 20μm×10μm, and the model adopts periodic boundary conditions, and the initial concentration of vacancies and fission gases in the entire simulated area is $c_v^{m,eq}$, $c_g^{m,eq}$. Calculated by Eq. (18), The diffusion coefficient at T=1200k is $D_3 = 8.72 \times 10^{-21}$ m^2/s.
T = 1276K, $0 < \tau < 550$s, the evolution of bubble distribution over time is shown in figure 2. The red area in the figure represents grains, while the blue area represents bubbles. At τ=155s as shown in figure 2 (a)-(b), neither model forms bubbles. At this time, vacancies and fission gases are dispersed in the grain. 20 seconds later, it can be observed that the model containing GB (figure 2 (d)), where GB absorb defects around them, and bubbles begin to nucleate. The distribution of defects in the grains becomes uneven, indicating the formation of nuclei. The model without GB on the left (figure 2 (c)) does not show this sign, and defects are still evenly dispersed in the grains. At τ =235s, it can be observed that both models have a large amount of nucleation. In the model without GB (figure 2 (e)), bubbles with closer distances fuse with each other to form larger bubbles, while another model (figure 2 (f)) has larger bubbles at the GB. The bubbles inside the grain have not yet fused but larger than those from the model without GB. In addition, obvious "bubble denuded zone" were observed near grain boundaries. When the simulation time τ reaches 500s, bubbles in both models have grown and fused. At this time, the bubbles distribution of the model without GB (figure 2 (g)) is similar to that in the grain of the model with GB (figure 2 (h)).
At τ = 235s, the model without GB has an average bubble diameter of 0.08μm, and the model with GB has a much larger intragranular bubbles diameter, which is 0.18μm. At τ = 500s, bubble diameter of the model without GB grows to 0.33μm, and the model with GB has a similar but smaller intragranular bubble diameter at this time, which is 0.31μm. It seems that in early stage, GB will promote the nucleation and growth of intragranular bubbles. After a period of simulation, the bubble sizes of the two models did not differ significantly. However, due to the fact that GB act as defects sink to absorb defects within the grains, the bubble sizes in the model containing GB are slightly smaller than those in the model without GB. Intergranular bubbles of the model with GB always have larger diameter, at τ = 235s, it is 0.25μm, and 0.43μm at τ = 500s. And it is also because GB act as defects sink, visible "void denuded zones" can be observed near GB. Width is defined as the average distance between the grain boundary and all bubbles closest to the grain boundary in this study. At τ = 235s, the width of the zone is 0.65μm, at τ = 500s the width of the zone is reduced to 0.49μm due to the growth of bubbles.

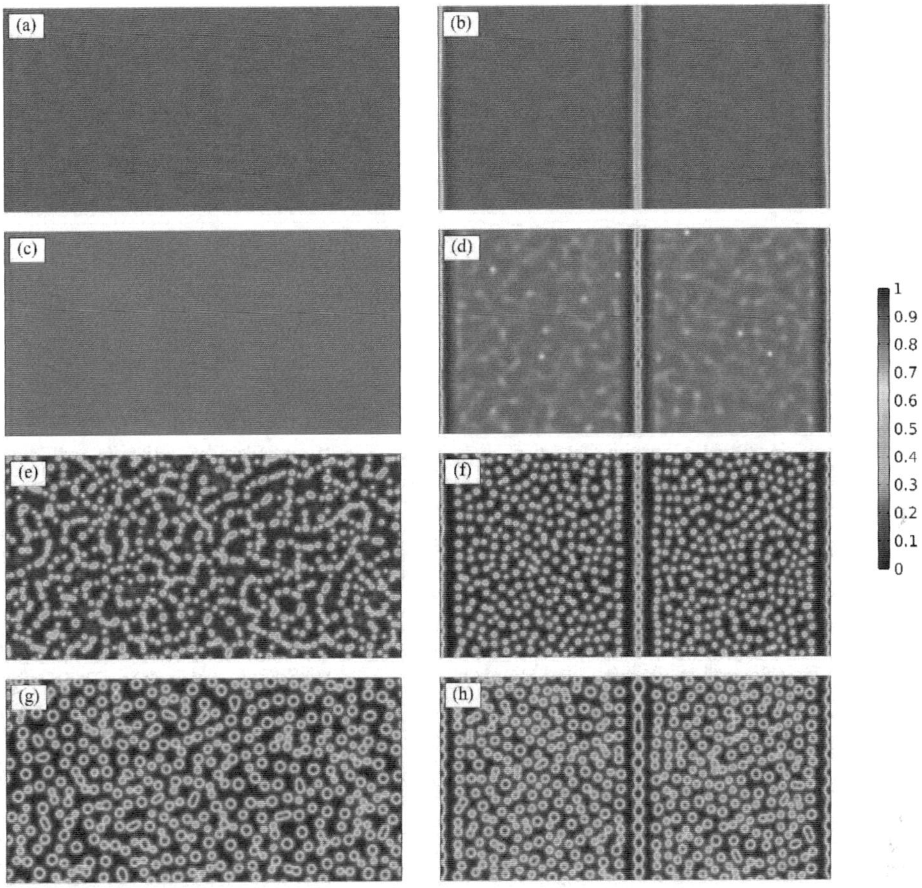

Figure 2. Distribution of gas bubbles evolution over time at T=1276K. (a)-(b) 155s; (c)-(d)175s; (e)-(f)235s; (g)-(h)500s. On the left is a model without GB, while the other side contains.

We further quantify the impact of GB on gas bubbles by using porosity. In our study, we defined porosity as the percentage of bubble area to the entire simulation area. In statistics, the grid points with $\eta \geq 0.8$ are considered as bubble phase. When T = 1276K, the variation of porosity with simulation time is shown in Figure 3 (a), including models without GB, models with GB, and models with GB but excluding areas with a width of 2μm near GB. As shown in Figure 3, regardless of the model, the evolution of porosity over time can be divided into three stages: the first stage is maintained at the zero level; the second stage is rapid growth stage; the third stage is the stage of slow and stable growth. Entering the second stage from the first stage signifies the beginning of nucleation, and entering the third stage indicates that no new bubbles are generated and existing bubbles continue to grow and fuse, the process of which is called Ostwald ripening[22]. These three stages are similar to those described in ref.[23].

From the figure 3 (a), it can be seen that the fission gas in the model with GB began to nucleate at τ = 165s, earlier than models without GB, which began to nucleate at 215s. Again, it indicates that the presence of GB promotes nucleation within grains. During the period from 180s to 250s, the porosity of the model with GB was always 0.005 higher than that of the model without GB. At τ = 200s, the porosity growth of models with GB begins to slow down, and the porosity of the model without GB maintains rapid growth, the gap in porosity between the two models is gradually narrowing. When the simulation reaches 270s, the porosity difference between the two models narrowed to 0.0008. There is almost no difference in porosity between the two models thereafter.

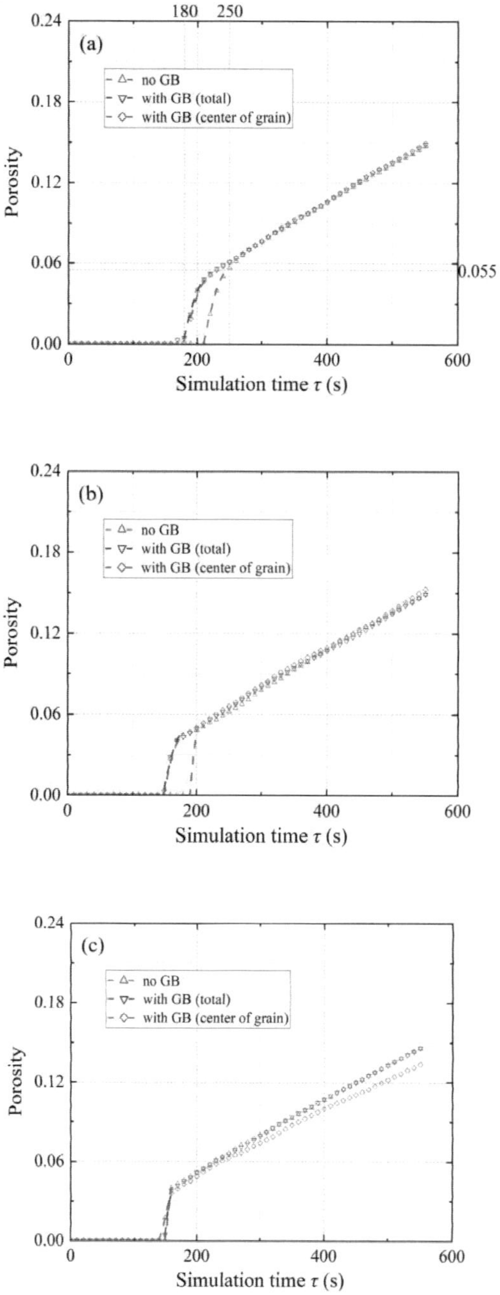

Figure 3. Variation of porosity with simulation time under three temperature conditions (a) T = 1276K; (b) T = 14276K; (a) T = 1676K

3.2. Effect of temperatures on nucleation

The effect of temperature on the release of fission gas will be discussed, and the influence caused by GB will not be considered this section. Two higher temperature conditions have been chosen: T= 1476K and 1676K. The diffusion coefficients under these two temperature conditions calculated by Eq. (19) and (20) are $D_2 = 1.583 \times 10^{-19}$ m^2/s, $D_1 = 5.476 \times 10^{-19}$ m^2/s, respectively.

The distribution of bubbles is shown in Figure 4. By comparing figure 2 (c) with figure 4 (a)(b), it can be observed that at τ = 175s, the fission gas at 1276K and 1476K has not yet formed bubbles, while visible bubbles have been formed at 1676K, with an average diameter of 0.35μm. At τ=175s, the bubble diameter at 1476K is 0.23μm, and the average diameter of bubbles at 1676K is 0.43μm. At τ=500s, the bubble diameter at 1476K is 0.38μm, and the average diameter of bubbles at 1676K is 0.56μm. The average diameter of bubbles formed under high temperature conditions at the same time is significantly larger than that formed at low temperature. and the number density of bubbles is lower than that at low temperature. Same situation as described in Li's[24] simulation. In addition, under T = 1276K, the nucleation of fission gas occurs in 215s, and 195s under T = 1476K, 150s under T = 1676K. The nucleation of fission gas under high temperature conditions occurs earlier than under low temperature conditions.

The variation of porosity with simulation time of the model without GB under T= 1276K, 1476K and 1676K is shown in Figure 5. Under high temperature conditions, the porosity of the model enters a growth state earlier, corresponding to earlier nucleation of bubbles. The increase in temperature leads to an increase in the diffusion coefficient of fission gases and vacancies. It is generally believed that a larger diffusion coefficient leads to the formation of larger bubbles in fission gas. When the simulation time reaches 300s, the difference in porosity of the model of three temperature conditions is small (<0.002). Considering that in our simulation, the source terms of vacancies and fission gases exist in the form of net generation terms, and the net generation term is a constant value. That is to say, the models under three temperature conditions have the same total concentration of fission gas and vacancies at the same time. So it is possible for models under three temperature conditions to have the same porosity at a certain time in our simulation. In addition, this is also related to our statistical method. When calculating the bubble phase, we consider all grid points with $\eta \geq 0.8$ as bubble phases. If the grid points with η >0.7 or 0.9 are considered bubble phases, the statistical results will be significantly different. But no matter what statistical method is used, the porosity of the model under high temperature conditions always enters the growth stage earlier than that under low temperature conditions.

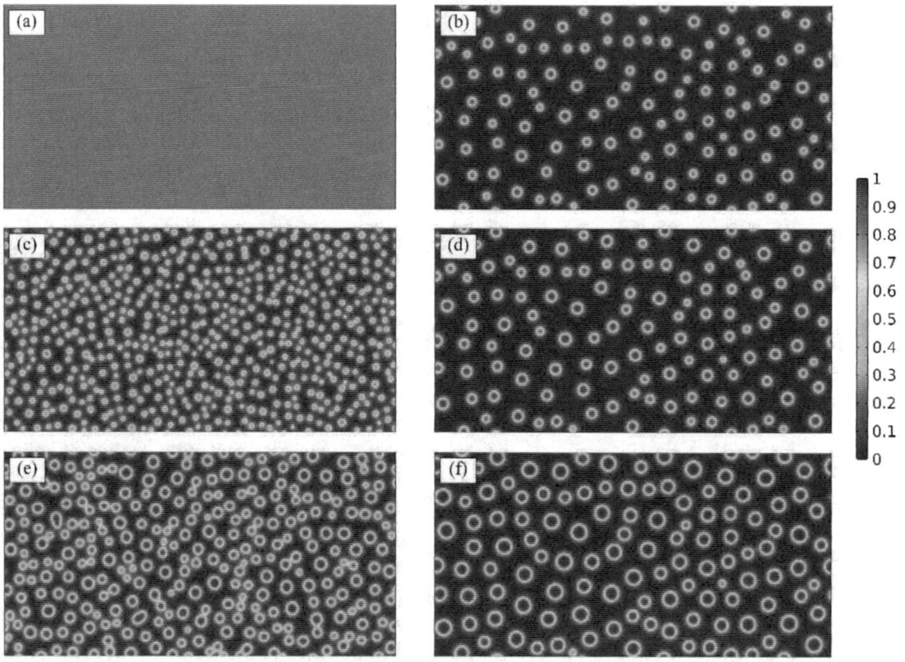

Figure 4. Distribution of bubbles under 1476K and 1676K in the model without GB (a) $T = 1476$K, $\tau = 175$s; (b) $T = 1676$K, $\tau = 175$s; (c) $T = 1476$K, $\tau = 235$s; (d) $T = 1676$K, $\tau = 235$s; (e) $T = 1476$K, $\tau = 500$s; (f) $T = 1676$K, $\tau = 500$s

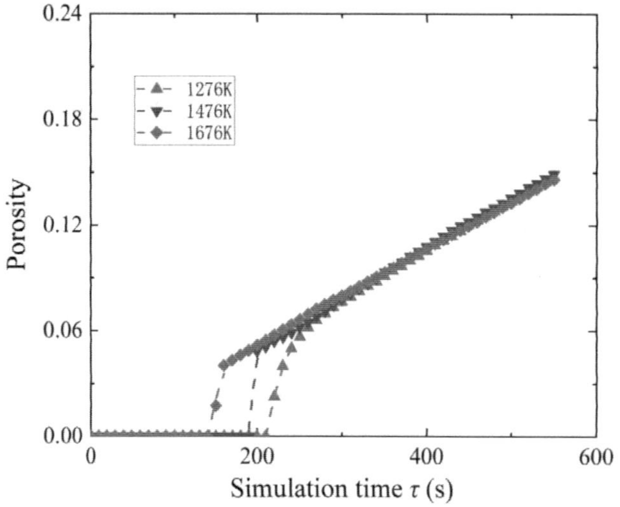

Figure 5. Variation of porosity with simulation time under three temperatures

3.3. Effect of grain boundaries at different temperatures

In section 3.1, it was found that GB have a promoting effect on the nucleation of fission gas bubbles. In section 3.2, it was found that high temperature conditions also promote the nucleation of bubbles, while promoting the formation of larger diameter bubbles. So, the impact of GB on the release of fission gases under high temperature conditions will be discussed in this section.

The variation of porosity with simulation time under T = 1476K and 1676K is shown in figure 3 (b)(c). It can be observed that the trend of porosity over time at 1476K is similar to that at 1276K, where models with GB nucleate earlier and porosity enters the growth state earlier. At T = 1476K, the fission gas in the model with GB began to nucleate at τ = 150s, not only earlier than the model without GB at T=1476K, but it also earlier than the model with GB at T=1276K, conforming to the conclusions in Sections 3.1 and 3.2. At T = 1476K, when the simulation time reaches a certain time, the difference in porosity between the two models is also very small which is similar to the situation at T=1276K.

But the situation is different at T=1676K. From Figure 3(c), it can be observed that the time difference between the two models when the porosity enters the growth stage is less than 5s, much smaller than at T=1276K and T=1476K. In the model with GB, nucleation occurs 5 seconds earlier at 1676K than at 1476K, the difference is not significant either. Considering that in this model, the diffusion coefficient of fission gas is the main factor affecting bubble nucleation, while the influence of GB on bubbles is reflected in the total free energy of GB. The total free energy varies very little with temperature, while the diffusion coefficient varies greatly with temperature, The total free energy at GB at 1676K is less than twice that at 1276K, but the diffusion coefficient is nearly 100 times that at 1276K. So it is normal for the variation of porosity over time at 1676K to differ from that at 1276K.

The distribution of bubbles is shown in Figure 6. With temperature increasing, the width of the "bubble denuded zone" increases. At τ = 235s, the width of the zone at T=1476K and 1676K are 1.03μm and 1.42μm, respectively. At τ = 500s, the width of the zone at T=1476K and 1676K are 0.89μm and 1.18μm, respectively. Different from Millett's work[23], no peak zones of bubbles were observed in our simulation. Because in our model, to simplify calculations, vacancies are generated by simulating the net generation term of vacancies. The recombination term and interstitial atom concentration were not considered. So there is no difference in diffusion coefficient between vacancies and interstitial atoms and no peak zones formed.

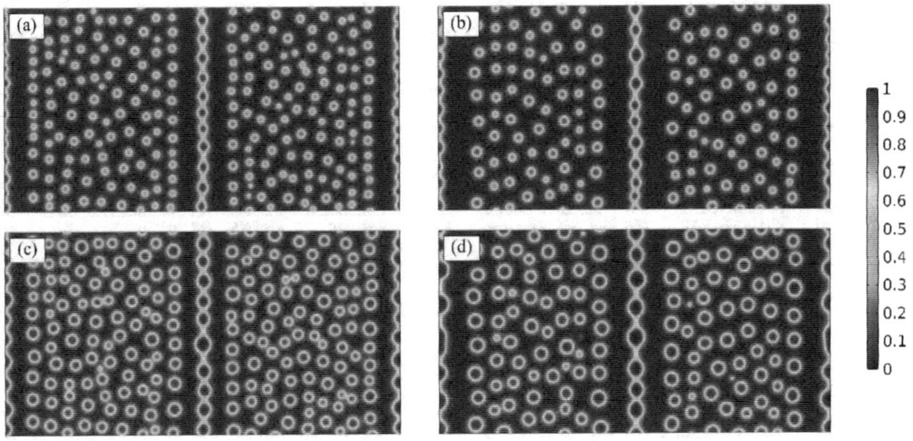

Figure 6. Distribution of bubbles under 1476K and 1676K in the model with GB (a) T = 1476K, τ = 235s; (b) T = 1676K, τ = 235s; (c) T = 1476K, τ = 500s; (d) T = 1676K, τ = 500s;

4. Conclusions

In this study, the fission gas diffusion coefficient given in the Turnbull model is used for establish a phase field model for the evolution of fission gas bubbles. The model uses parabolic approximation of free energy to simplify calculations and improve convergency. By establishing the models with and without GB, the influence of GB on the formation of bubbles from fission gas and the degree of influence at different temperatures have been studied.

Research has shown that GB have a promoting effect on the nucleation of fission gas bubbles within grains, and the lower the temperature, the stronger the promoting effect. The promoting effect under high temperature conditions is not significant. In addition, the model also demonstrates the influence of temperature on bubble nucleation. The higher the temperature, the larger the diffusion coefficient, and the larger the bubble size and the smaller the number density at the same time. The "bubble denuded zone" also widens with increasing temperature.

The model also represents the three stages of the formation of fission gas bubbles: incubation stage, nucleation stage, and growth stage. During the incubation stage, no bubbles are generated and the porosity does not increase; During the nucleation stage, a large number of bubbles form and preferentially form at grain boundaries, resulting in a rapid increase in porosity; During the growth stage, bubbles stop producing, existing bubbles continue to grow and fuse with adjacent bubbles, resulting in a stable increase in porosity.

Acknowledgments

The authors acknowledge the support from the Fundamental Research Funds for the Central Universities (3072022JC1502) of Harbin Engineering University.

References

[1] Rest J 2010 *J. Nucl. Mater.* **402** 179
[2] Trinkaus H, Singh B N 2003 *J. Nucl. Mater.* **323** 229
[3] Rest J, Hofman G L 1999 *Nucl. Technol.* **126** 88
[4] Pastore G, Luzzi L, Di Marcello V, Van Uffelen P 2013 *Nucl. Eng. Des.* **256** 75
[5] Piro M H, Sunderland D, Livingstone S, Sercombe J, Revie R W, Quastel A, Terrani K A, Judge C 2020 Comprehensive Nuclear Materials (2nd Edition) (Amsterdam: Elsevier) p248
[6] He W, Wu X Y, Wu L, Wen B, Zhu W, Zhang W, Pan R J, Wang Z, Huang W J 2017 *Nucl. Power Eng.* **38** 170
[7] Moelans N. Blanpain B. Wollants P 2008 *Comput. Coupling Phase Diagrams Thermochem* **32** 268
[8] Guo C, Kang C R, Gao Y, Zhang Y C, Deng G Y, Ma C, Xu C J, Liang S H 2022 *Acta Phys. Sin.* **71** 263
[9] Hu S, Henager Jr C H 2009 J. *Nucl. Mater.* **394** 155
[10] Millett P C, El-Azab A, Rokkam S, Tonks M, Wolf D 2011 *Comput. Mater. Sci.* **50** 949
[11] Aagesen L K, Schwen D, Tonks M R, Zhang Y 2019 *Comput. Mater. Sci.* **161** 35
[12] Jiang Y B, Liu W B, Li W J, Sun Z Y, Xin Y, Chen P H, Yun D 2021 *Comput. Mater. Sci.* 188 110176.
[13] Zhao J J, Sun D, Xi L, Chen P, Zhao J J, Wang Y Y 2023 *Phys. Chem. Chem. Phys.* 25 14928
[14] Cahn J W, Hilliard J E. 1958 J. Chem. Phys. **28** 258
[15] Li Y, Hu S, Montgomery R, Gao F, Sun X 2013 *Nucl. Instrum. Methods Phys. Res., Sect. B* **303** 62
[16] Cahn J W, Hilliard J E. 1958 *J. Chem. Phys.* **28** 258
[17] Cahn J W 1961 *Acta Metall.* **9** 795
[18] Allen S M, Cahn J W 1972 *Acta Metall.* **20** 423
[19] Allen S M, Cahn J W 1973 *Scr. Metall.* **7** 1261
[20] Turnbull J A, Friskney C A, Findlay J R, Johnson F A, Walter A J 1982 *J. Nucl. Mater.* **107** 168
[21] Turnbull J A, White R J, Wise C 1989 Technical committee on water reactor fuel element computer modelling in steady state, transient and accident conditions Preston, United Kingdom, September 18-22, 1988 p174
[22] Sagui C, Grant M 1999 *Phys. Rev. E* **59** 4175.
[23] Millett P C, El-Azab A, Wolf D 2011 *Comput. Mater. Sci.* **50** 960
[24] Li Y L, Hu S Y, Sun X, Gao F, Henager C H, Khaleel M 2011 *Sci. China: Phys., Mech. Astron.* **54** 856.

The compatibility of nitrided W18Cr4V steel in a sodium environment

Xiaogang Fu, Na Liang, Wei Zhang, Liu Tao, Bo Qin, Zhangshun Ruan, Bin Long

China Institute of Atomic Energy, Xinzhen, Fangshan Dist, Beijing, China

E-mail: binlong@ciae.ac.cn

Abstract. The guideway friction pairs are the key components of the loading elevator and unloading elevator in the refueling system for sodium cooled fast reactor. With the excellent hardness and wear resistance in air, nitrided W18Cr4V steel is a candidate material for the guideway friction pairs. In order to evaluate the applicability in the high-temperature sodium, it is important to conduct the compatibility tests of nitrided W18Cr4V steel for different exposure time. The results show that the nitrogen content of nitrided W18Cr4V steel decreased obviously after 1000 h, 2000 h and 3000 h sodium compatibility tests at 540 ℃ and the nitrided strengthening effect of W18Cr4V steel gradually decreased with the test duration increasing.

1. Introduction

The refueling system for sodium cooled fast reactor is mainly composed of in-reactor refueling system, out-of-reactor refueling system, new fuel assembly transfer system and spent fuel assembly transfer system. The loading elevator and unloading elevator are the main equipment of the refueling system which are used for the fuel assembly transfer operation. Considering the operation reliability of loading elevator and unloading elevator, guideway friction pairs are the key components composing of a guide rail and a slide block[1-3]. Nitrided W18Cr4V steel is a candidate material for the guideway friction pairs of the elevators, which has the potential to meet the performance requirements in a sodium environment.

Nitriding treatment is an industrial manufacturing method widely used to improve the surface properties of steels. Through gas nitriding or ion nitriding, a nitrided layer can be formed on the surface of the steels with high hardness and high wear resistance[4-7]. Although nitrided steels exhibit excellent performance in air environment, there are relatively few studies regarding their behavior and performance evaluation in sodium at a high temperature[8-13]. In order to confirm the application feasibility of the guideway friction pairs, nitrided W18Cr4V steel has been tested in a high temperature sodium environment with different exposure times to determine the nitrogen diffusion behavior and compatibility performance change.

2. Experimental procedures

The compatibility tests were carried out in static sodium facility. The size of nitrided W18Cr4V steel specimens were 20 mm × 50 mm × 5 mm. The specimens were capsuled into the different claddings of 304 stainless steel. The specimen was immersed in the sodium and the gap was filled with high purity Ar gas. The compatibility tests were conducted at 540℃ for 1000 h, 2000 h and 3000 h. After the compatibility tests, the specimens were cleaned in alcohol at room temperature to remove the residual sodium on the surface. When the test specimens were placed in alcohol, bubbles were formed on the surface of the specimens and moved to the surface of alcohol. When the bubbles were no longer generated, the residual sodium had been removed completely. Generally, the cleaning time in alcohol varied from 2 hours to 3 hours due to the difference of the residual sodium quality. Next, the specimens were washed twice in high purity water for 5 minutes each by ultrasonic cleaning machine. The water on the surface of the specimens was dried with cold air. Finally, the specimens were placed in an drying cabinet at 60°C for one hour. Then the macroscopic morphology of specimens was observed using an optical camera. The mass change of specimens was determined by the analytical balance. Part of the specimens were cut into two pieces and the cross section was prepared by mechanical grinding and polishing. A scanning electron microscope (SEM) with energy dispersion X-ray (EDX) was used to characterize nitrogen content of specimens. The microhardness on the cross section of specimens was measured by a microhardness tester.

3. Results and discussion

Figure 1 shows SEM image of cross section for original nitrided W18Cr4V steel after microhardness test. There are five microhardness indentation points from the specimen surface to the substrate with the interval of 30-40 μm. It can be seen that the microhardness indentation area increases with the distance increasing away from the specimen surface. It indicates that the microhardness reduces gradually with the distance increasing away from the surface at the cross section of original specimen.

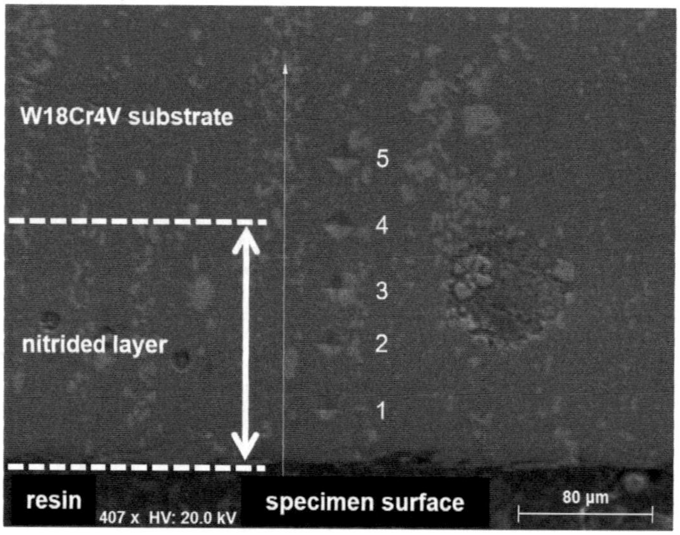

Figure 1. SEM image of cross section for original nitrided W18Cr4V steel

The microhardness and results at the cross section of original nitrided W18Cr4V steel are given in Figure 2. The microhardness at the cross section of original specimen is reduced from 1212.5$Hv_{0.1}$ to 846.6$Hv_{0.1}$.

The microhardness of original specimen substrate is 861.4$Hv_{0.1}$. The thickness of nitrided layer for original specimen is about 140 μm determined by microhardness change at the cross section.

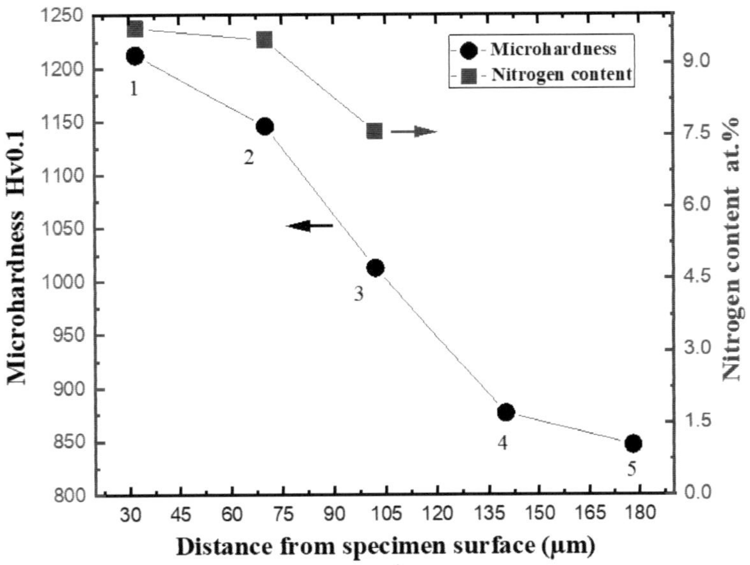

Figure 2. The microhardness and nitrogen content results at the cross section of original nitrided W18Cr4V steel

Figure 2 also shows the results of EDX analysis for original nitrided W18Cr4V steel. EDX analysis was performed at the parallel positions of microhardness indentation points of the original specimen shown in Figure 1 and the nitrogen content at the corresponding positions were obtained. The results indicate the microhardness in the nitrided layer is related with the nitrogen content. When the nitrogen content in the nitrided layer gradually decreased, the microhardness value reduced correspondingly. It means that the surface strengthening effect of W18Cr4V steel is positively correlated with the nitrogen content in the nitrided layer.

Figure 3 shows the optical macrographs of nitrided W18Cr4V steel before and after 1000 h, 2000 h and 3000 h compatibility tests in sodium at 540 ℃. The specimens had no black spots after the compatibility tests in sodium. But when the specimens were placed in alcohol, some black spots appeared on the surface. When the specimens were washed in high purity water, some black spots continued to occur. They were the thin black oxide films formed on the surface of specimens, indicating that the surface corrosion resistance of the nitrided W18Cr4V steel decreased after the high-temperature sodium tests.

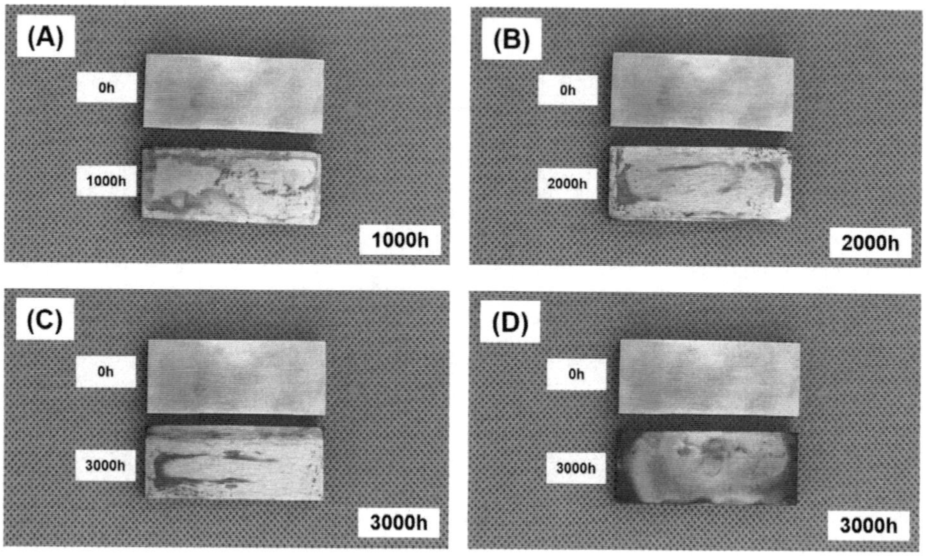

Figure 3. The optical macrographs of nitrided W18Cr4V steel before and after tests in sodium at 540 ℃ (A) 1000 h, (B) 2000 h, (C) 3000 h, (D) 3000 h.

The mass change results of nitrided W18Cr4V steel before and after 1000 h, 2000 h and 3000 h compatibility tests in sodium at 540℃are given in Table 1. It shows that the mass of nitrided W18Cr4V steel specimens decreased after tests in sodium at 540℃. These data indicate that some elements in the nitrided W18Cr4V steel dissolved and diffused into the high-temperature sodium. It resulted in mass loss of nitrided W18Cr4V steel specimens after the compatibility tests and also caused the decrease of surface corrosion resistance as shown in Figure 3.

Table 1. The mass change results of nitrided W18Cr4V steel specimens before and after tests in sodium at 540℃

Test Time (h)	Mass before Test (g)	Mass after Test (g)	Mass Loss (g)
1000	43.2703	43.2389	0.0314
2000	43.3527	43.3218	0.0309
3000	43.3715	43.3377	0.0338
3000	43.2917	43.2620	0.0297

Figure 4 shows the microhardness results of nitrided W18Cr4V steel specimens before and after 1000 h, 2000 h and 3000 h compatibility tests in sodium at 540 ℃. The average microhardness in nitrided layer of specimens decreased from 1123.6$HV_{0.1}$ to 512.2$HV_{0.1}$ - 589.6$HV_{0.1}$ after sodium tests at 540℃. At the same time, the average microhardness of nitrided W18Cr4V steel substrate decreased from 861.4$HV_{0.1}$ to 470.2$HV_{0.1}$ - 558.7$HV_{0.1}$. It indicates the microhardness of nitrided layer and substrate have the similar decrease trend after tests in sodium at 540 ℃. After 1000 h test in sodium at 540℃, the average microhardness of nitrided layer decreased by 534$HV_{0.1}$, while the average microhardness of substrate decreased by 302.7$HV_{0.1}$. The microhardness reduction of nitrided layer was

2749 (2024) 012002 doi:10.1088/1742-6596/2749/1/012002

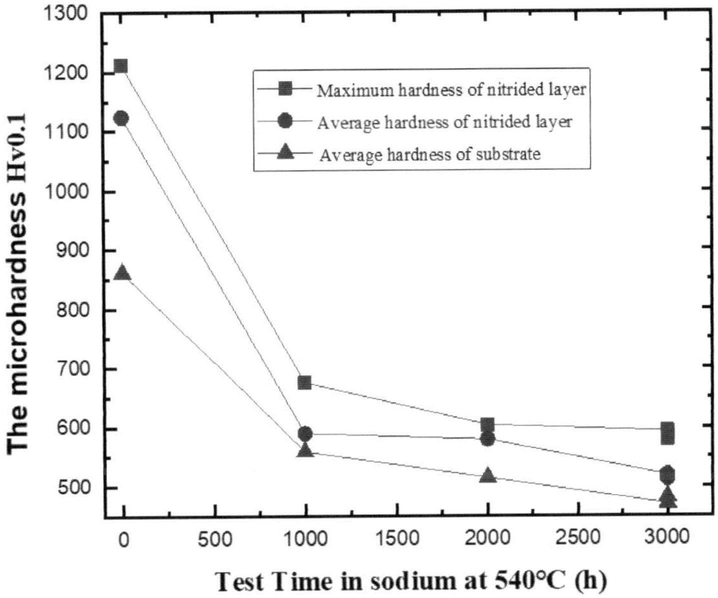

Figure 4. The microhardness results of nitrided W18Cr4V steel before and after different tests in sodium at 540℃

significantly higher than that of substrate. After 2000 h and 3000 h sodium tests at 540℃, the average microhardness differences between nitrided layer and substrate are 64.2$HV_{0.1}$ and 39.7$HV_{0.1}$ respectively. The microhardness of nitrided layer and substrate are gradually close to each other.

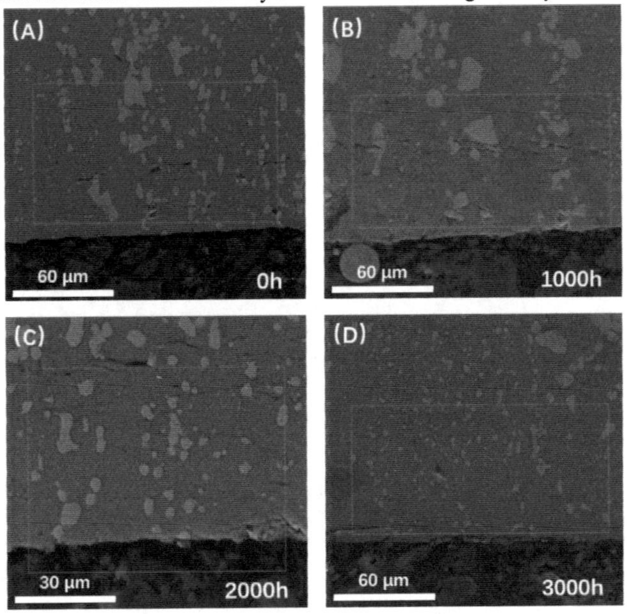

Figure 5. The SEM images of the cross section of nitrided W18Cr4V steel before and after tests in sodium at 540 ℃ (A) 0 h, (B) 1000 h, (C) 2000 h, (D) 3000 h.

Figure 5 shows the SEM analysis results of the cross section for nitrided W18Cr4V steel specimens before and after 1000 h, 2000 h and 3000 h test in sodium at 540 ℃. There are some white tungsten carbide particles and a certain number of cracks on the surface of the specimens. These cracks in the original specimen may had been generated during the nitriding treatment process. Nitrogen atoms located in the interstitial position were used to increase the surface hardness of W18Cr4V steel. In order to observe the strengthening effect in sodium, a certain area of the nitriding layer was analysed by EDX.

Figure 6 shows the EDX analysis results of the cross section of nitrided W18Cr4V steel specimens before and after 1000 h, 2000 h and 3000 h test in sodium at 540 ℃. In Figure 3(A) the EDX analysis result of original specimen has an obvious nitrogen peak near 0.39keV. However, nitrogen peaks gradually disappear after 1000 h, 2000 h and 3000 h sodium tests at 540 ℃. The thickness of nitrided layer decreased from 140 μm to near zero after sodium tests at 540℃. It indicates that a large amount of nitrogen in the nitrided layer had diffused into the liquid sodium at 540 ℃. It resulted in a mass loss of nitrided W18Cr4V steel specimen and a decrease of corrosion resistance, microhardness and nitrogen content in the nitrided layer. Considering the movement trend of nitrogen in the nitrided layer, the surface nitrided strengthening effect of W18Cr4V steel gradually decreased with the sodium test duration increasing and nitrided W18Cr4V steel may not be able to maintain the sufficient hardness and wear resistance in a sodium environment at the high temperature.

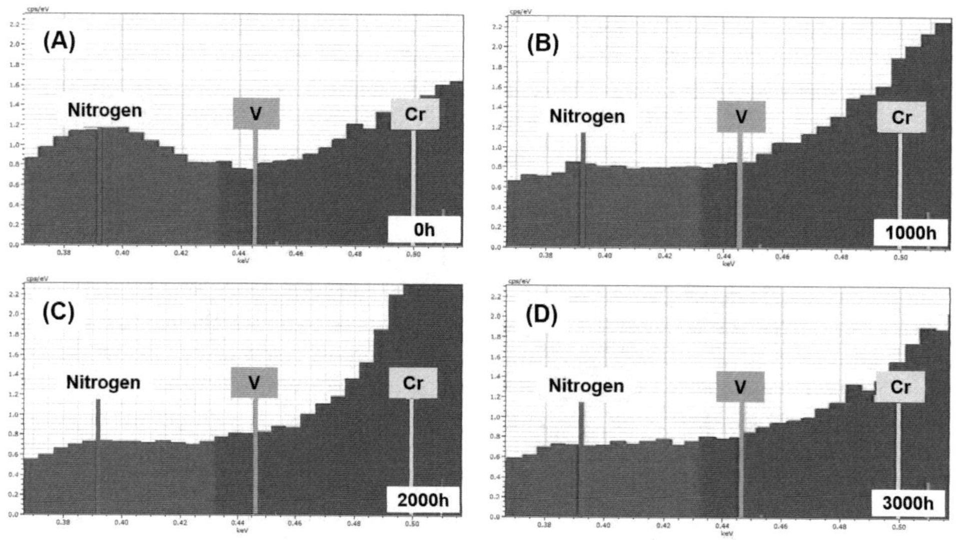

Figure 6. The EDX analysis of the cross section of nitrided W18Cr4V steel before and after tests in sodium at 540 ℃ (A) 0 h, (B) 1000 h, (C) 2000 h, (D) 3000 h.

4. Conclusions

The compatibility tests of nitrided W18Cr4V steel were conducted in sodium at 540 ℃ for 1000 h, 2000 h and 3000 h. Based on the results discussed in previous section, conclusions were summarized in following:

1. Nitrogen in the nitrided layer diffused into the liquid sodium at 540 ℃ obviously after sodium compatibility tests at 540 ℃.

2. Nitrogen lost to sodium resulted in a mass reduction of specimens and a decrease of corrosion resistance, microhardness and nitrogen content in the nitrided layer.

3. Due to nitrogen diffusing into liquid sodium at 540 °C, nitrided strengthening effect of W18Cr4V steel may be difficult to maintain for a long time.

5. References

[1] Tashlykov O L, Dolgii Yu F and Sesekin A N. 2021 Optimization of refueling times in fast neutron reactors[J]. *IOP Conf. Series: Materials Science and Engineering.* **1089** 012006.

[2] Dushev S A, Timofeev A V, Lyubimov M A, and Belinskii D L. 2020 Refueling system for fast reactors: Creation and development[J]. *Atomic Energy,* **129**(2) 87-92.

[3] Wang M Z, Dong S G, Song Q, Yang K L, Gu J P, Yan H, Wang C L, Yu T J, Jin F L, Ma H S, Zhu H and Zhao L X. 2008 The verifying test of refueling system of the China Experimental Fast Reactor[J]. *Chinese Journal of Nuclear Science and Engineering,* **28**(3) 210-217.

[4] Xi Y T, Liu D X, Han D, Han Z F. 2008 Improvement of mechanical properties of martensitic stainless steel by plasma nitriding at low temperature[J]. *Acta Metall. Sin.(Engl. Lett.),* **21**(1) 21-29.

[5] Gerasimov S A, Zhikharev A V, Berezina E V, Zubarev G I and Pryanichnikov V A. 2004 New ideas on the mechanism of structure formation in nitrided steels[J]. *Metal Science and Heat Treatment,* **46**(1) 13-17.

[6] Karami M B, Gercekcioglu E. 2000 Wear behaviour of plasma nitrided steels at ambient and elevated temperatures[J]. *Wear,* **243** 76-84.

[7] Emami M, Ghasemi H M and Rassizadehghan J. 2010 High temperature tribological behaviour of 31CrMoV9 gas nitrided steel[J]. *Surface Engineering,* **26**(3) 168-172.

[8] Johnson R N, Schrock S L, and Whitlow G A. 1974 Wear resistant coatings for reactor components in liquid sodium environments[J]. *J. Vac. Sci. Technol,* **11**(4) 759-764.

[9] Johnson R N. 1984 Coatings for fast breeder reactor components[J]. *Thin solid film,* **118** 31-47.

[10] Hemant Kumar, Ramakrishnan V, Albert S K, Bhaduri A K and Ray K K. 2017 Friction and wear behaviour of Ni-Cr-B hardface coating on 316LN stainless steel in liquid sodium at elevated temperature. *Journal of Nuclear Materials,* **495** 431-437.

[11] Natesan K. 1983 Influence of nonmetallic elements on the compatibility of structural materials with liquid alkali metals[J]. *Journal of Nuclear Materials,* **115** 251-262.

[12] Furukawa T, Kato S and Yoshida E. 2009 Compatibility of FBR materials with sodium[J]. *Journal of nuclear material,* **392** 249-254.

[13] Fidler R S and Collins M J. 1965 A review of corrosion and mass transport in liquid sodium and the effects on the mechanical properties. *Atomic Energy Review,* 3-50.

Acknowledgments

Authors wishing to acknowledge senior engineer Haoran Ma for the microhardness test support and the invaluable discussion.

Ligin-derived graphene-like ultrathin carbon materials for ORR catalysis

Wenting Liu[1,2], **Xiaoyu Wu**[1 §]
[1] School of Automobile and Transportation, Shenzhen Polytechnic University, Shenzhen 518055, China
[2] School of Materials Science and Engineering & Guangxi Key Laboratory of Information Materials, Guilin University of Electronic Technology, Guilin 541004, China

§ Corresponding author. E-mail: wuxiaoyu@szpu.edu.cn

Abstract. Low-cost nitrogen-doped carbon materials with graphene-like structures were designed and prepared from extensive renewable biomass waste lignin. The obtained carbon materials exhibited well-designed ultrathin two-dimensional structures with high specific area and mesoporous characteristics. Benefiting from the well-designed structures with extensive catalytic active sites and efficient mass transfer and diffusion, as well as the modified electronic characteristics from nitrogen doping, the lignin-derived carbon materials exhibited pronounced ORR catalytic activity and methanol tolerance. Considering these, the lignin-derived carbon materials hold great application potentials as ORR catalysts and catalyst supports.

1. Introduction

Hydrogen energy, as a typical green secondary energy, has been gradually becoming a important carrier of global energy transformation and development. Fuel cell is one of the key technologies of hydrogen energy application. Due to its superiorities in high energy conversion efficiency, clean process, almost zero emission, long life and wide range of hydrogen fuel sources, it holds great application prospects in important scientific and technological fields such as vehicles, drones, robots and communication stations. As the core material of hydrogen fuel cell, the cost of precious metal platinum catalyst accounts for about 40% of the reactor system, which seriously limits its large-scale commercial application. Therefore, in order to cope with the development of hydrogen energy industry, achieve breakthroughs in hydrogen fuel cell technology, and comply with the development trend of high performance, multi-function and green development of new materials, designing and developing high-performance non-noble metal catalysts is the key topic and key technology of the current hydrogen energy industry and hydrogen fuel cell research, which has important theoretical research and practical application significance.

Due to the large specific surface area, rapid ion diffusion and expanded surface active sites, two-dimensional materials have attracted extensive attention in electrochemical applications, especially catalysis. Recently, graphene-like nitrogen-doped carbon ultrathin two-dimensional materials are regarded as a new generation of high-performance oxygen reduction reaction (ORR) catalysts that are expected to replace traditional precious metal platinum catalysts [1-3]. At the same time, as a catalyst carrier, the large specific surface area of graphene-like nitrogen-doped carbon ultrathin two-dimensional materials promotes high density exposure of catalytic active sites and high conductivity advantages, which also provides more possibilities for designing non-noble metal catalysts. In order to avoid the complex preparation procedure of traditional nitrogen-doped carbon materials and dependence on toxic raw materials, the use of low-cost, widely sourced, environmentally friendly and renewable biomass raw materials for derivative preparation is regarded as a promising alternative [4]. Lignin is the second largest reserves of natural biomass polymer on the earth, with a unique polymer structure and rich functional group characteristics (phenolic polymer functional group, phenyl skeleton, unique C3 chain structure and highly compact bonding form). It is a high-quality precursor for preparing graphene-like ultrathin carbon, nitrogen-doped carbon and other high-performance carbon materials with excellent physical and mechanical strength and chemical activity [5].

Based on these considerations, nitrogen-doped carbon materials with graphene-like structures were designed and prepared based on extensive renewable biomass waste lignin, which showed good ORR catalytic activity and had great potential in the application of catalysts and catalyst carriers for hydrogen fuel cell cathode.

2. Experimental details

2.1. Materials and methods

The raw material of lignin powder was carefully purified in aqueous solution of 20 wt % KOH and 1 M HCl in sequence to remove the residual inorganic species. To prepare carbon materials, 0.2g lignin and 0.2 g Ni powders were mixed uniformly in agate mortar, then calcine the mixed powders (in porcelain boat) in tube furnace with controlled temperature and gas atmosphere step by step: 1) 400°C for 2h (heating rate: 10 °C min^{-1}) with 100 sccm He gas, 2) 600°C for 2h (heating rate: 10 °C min^{-1}) with 50 sccm He gas, 3) 600°C for 2h (heating rate: 10 °C min^{-1}) with 50 sccm He gas, 4) 1000°C for 0.5 h (heating rate: 20 °C min^{-1}) with 50 sccm He gas, 5) rapidly cool to room mate with 50 sccm He gas. Finally, the obtained powders were added into 2M $Fe(NO_3)_3$ aqueous solution, stirred at 80°C for 12h, then the final solid products were filtered and dried at 45°C overnight.

2.2. Characterization

X-ray diffraction (XRD, Rigaku D/max-2200PC) was used to characterize the crystallographic structure of the lignin-derived product. The microstructure and detailed morphology was observed using field-emission scanning electron microscope (SEM, JEOL JSM-7500F) and transmission electron microscope (TEM, FEI Gatan F30). Brunauer-Emmett-Teller (BET) measurements were performed on an ASAP-2010 analyzer. X-ray photoelectron spectroscope (XPS, AXIS UTLTRADLD) was used to analyze the elemental composition. A Reishaw Microscope System (RM2000) was used for Raman measurements (at 532 nm).

2.3. Performance measurements

PARSTAT 2273 electrochemistry system was used for electrochemical performance measurements of the product with a conventional three-electrode electrochemical cell. The catalysts-loaded glassy carbon rotating disk electrode was used as the working electrode, graphite rod was selected as counter electrode, the reference electrode was Ag/AgCl electrode, and the electrolyte was 0.1 M KOH aqueous solution. All electrochemical measurements were conducted at room temperature.

3. Results and discussion

Lignin exhibits a typical structure of aromatic rings with rich functional group characteristics (phenolic polymer functional group, phenyl skeleton, unique C3 chain structure and highly compact bonding form), which can serve as a potential precursor for nitrogen-doped carbon synthesis. Fig. 1a presents the XRD pattern of the obtained carbon materials, which presents two broad characteristic peaks at ~23° and 44.0°, as well as a weaker and broader peak at ~15°. The low-intensity characteristic peaks can be indexed to the (002) and (100) planes of graphite with low degree of graphitization [6]. And the weak peak at ~15° indicates the existence of N incorporation. According to literatures, nitrogen doping can modify the electronic characteristics of carbon and modulate the local charge density distribution to realize superior ORR activity [7, 8].

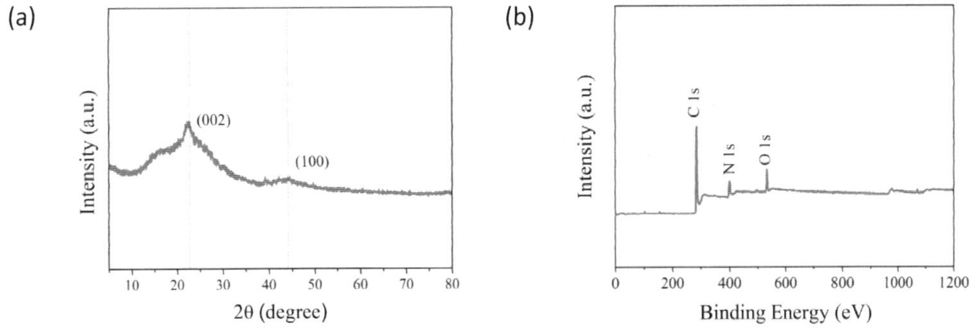

Fig. 1 (a) XRD pattern and (b) XPS survey spectra of the lignin-derived nitrogen-doped carbon.

To further verify the existence of N incorporation in the product, XPS measurements were conducted. According to the XPS survey spectra in Fig.1b, the presence of C, N and O was observed in the lignin-derived product, thus confirming the composition as N-doping carbon [9]. While the existence of O element may be ascribed to the physicochemical absorbed oxygen and the unavoidable oxidation reactions of the product when exposed to the air [10]. Meanwhile, Raman spectrum was also obtained to further reveal the composition and structure features. As shown in Fig. 2, the G band (~1580 cm^{-1}) refers to the E_{2g} vibration mode of graphitic sp^2 carbon domain, which further confirms the graphene-like structures of this lignin-derived product. While the obvious D band (~1360 cm^{-1}) with high density of I_D/I_G~0.92 corresponds to the partially disordered structures, indicating the existence of massive structural defects caused by nitrogen doping [11].

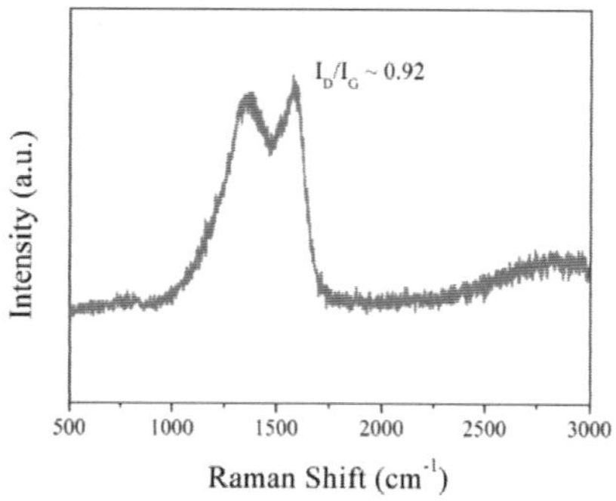

Fig. 2 Raman spectrum of the lignin-derived nitrogen-doped carbon.

The detailed morphology and microstructure of the lignin-derived nitrogen-doped carbon were investigated by SEM and TEM. As shown in the SEM image in Fig. 3, the obtained carbon materials exhibit obvious morphology of interconnected ultrathin nanosheets, which are almost transparent and curved. More clearly, according to the TEM image in Fig. 4a, the obtained product are assembled with two-dimensional graphene-like ultrathin flexible nanosheets. As shown in the magnified TEM image in Fig. 4b, the ultrathin flexible nanosheets exhibit almost transparent morphology and shows obvious wrinkles or corrugation. Meanwhile, according to the further-magnified TEM image of Fig. 4c, obvious porous structures of ~2 nm in size were observed in the surface of the flexible nanosheets, suggesting their mesoporous characteristics.

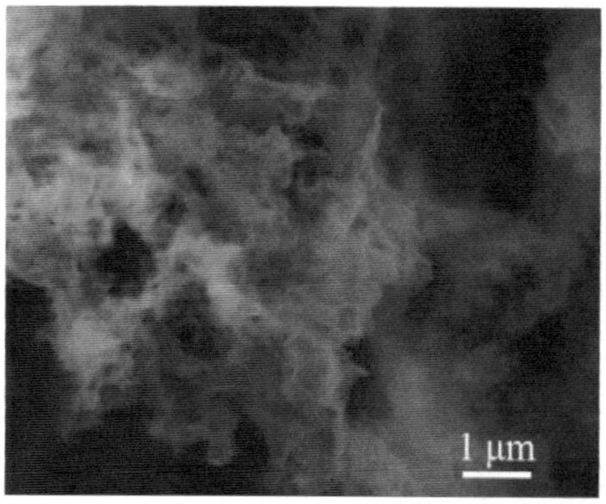

Fig. 3 SEM images of the lignin-derived nitrogen-doped carbon

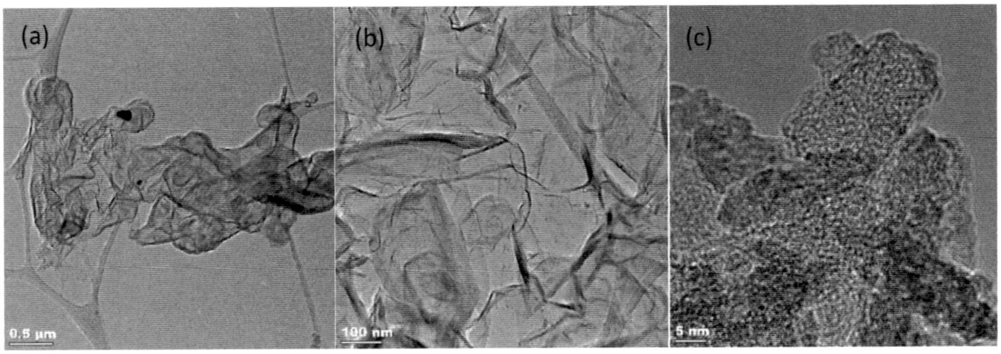

Fig. 4 TEM images of the lignin-derived nitrogen-doped carbon.

In order to calculate the specific surface area and confirm the porous features of the obtained carbon materials, full nitrogen adsorption/desorption isotherms were also collected. As shown in Fig. 5a, a type IV curve with a H3-type hysteresis loop was observed clearly in the isotherm, further indicating the mesoporous structures existed in the graphene-like carbon materials [12]. Meanwhile, according to the adsorption data, the specific surface area of the the carbon materials can be calculated to be 289.37 $m^2 g^{-1}$. Furthermore, as shown in Fig. 5b, obvious peaks at the range of 2-4 nm can be observed in the pore size distribution pattern, corresponding to the mesoporous channels in the product. These results are consistent with the TEM images discussed above.

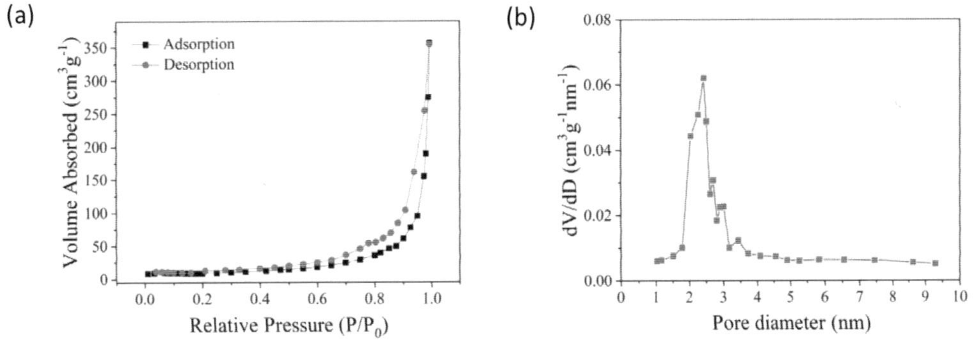

Fig. 5 (a) N_2 adsorption/desorption isotherms and (b) the pore size distribution of the lignin-derived nitrogen-doped carbon.

Considering the graphene-like structure with high specific area and mesoporous characteristics, and the advantageous composition with nitrogen-doping, the lignin-derived nitrogen-doped carbon materials are expected to be promising in ORR catalysis. To confirm the catalytic activities of the lignin-derived carbon materials, cyclic voltammetry (CV) was performed at a scan rate of 50 mV s^{-1}. As shown in Fig. 6a, the obtained nitrogen-doped carbon exhibits an obvious cathodic peak in O_2-saturated electrolyte of 0.1 M KOH aqueous solution, thus confirming its ORR catalytic activities. In contrast, in N_2-saturated electrolyte, featureless peak currents are observed in the CV curve of our products. The linear sweep voltammetry (LSV) curves obtained using the RDE technique at a rotation rate of 1600 rpm in Fig. 6b also verify the considerable ORR catalytic performance, although not comparable to the commercial Pt/C catalyst in both onset potentials and current densities. Moreover, when adding 3.0 M methanol into the O_2-saturated electrolyte, no obvious changes can be observed in the CV curve of the lignin-derived nitrogen-doped carbon, suggesting its excellent tolerance to methanol. Moreover, as shown in Fig.7, in long term test, the lignin-derived nitrogen-doped carbon exhibit remarkable stability in contrast to commercial Pt/C catalyst. Only slight current attenuation can be observed in the ORR current-time chronoamperometric response curve of the lignin-derived nitrogen-doped carbon when adding 3.0 M methanol into the O_2-saturated electrolyte. In fact, this methanol tolerance and catalytic stability is of great significance for its applications in practical hydrogen fuel cells.

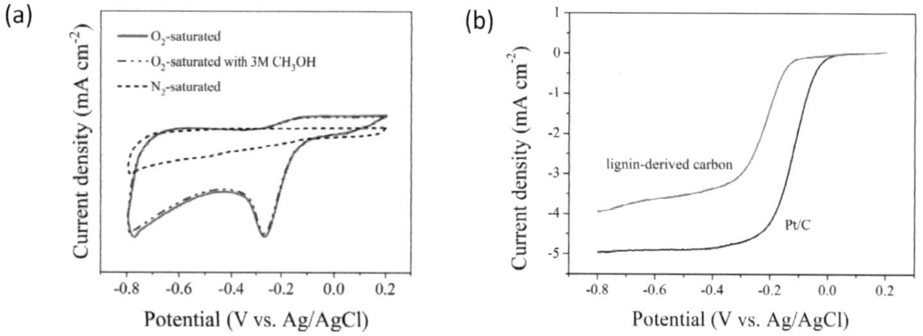

Fig. 6 (a) CV curves of the lignin-derived nitrogen-doped carbon in O_2-saturated electrolyte, N_2-saturated electrolyte and O_2-saturated electrolyte with 3.0 M CH_3OH; (b) LSV curves of the lignin-derived nitrogen-doped carbon and Pt/C in O_2-saturated electrolyte at a rotation rate of 1600 rpm.

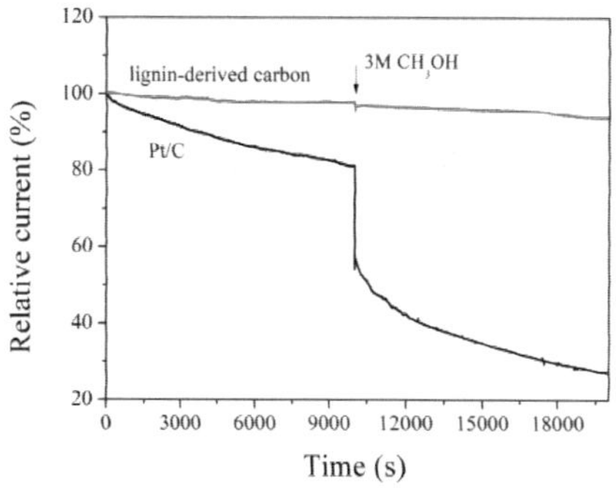

Fig. 7 Current-time chronoamperometric response of the lignin-derived nitrogen-doped carbon and Pt/C in O_2-saturated electrolyte at a rotation rate of 1600 rpm, in which the arrow indicate the introduction of 3.0 M methanol.

4. Conclusions

With the addition of Ni metal as structural template and a facile pyrolysis procedure of biomass waste lignin, we successfully synthesized a low-cost and renewable carbon material. Characterization results confirmed that the obtained carbon material exhibited remarkable graphene-like two-dimensional structures, with a specific surface area of 289.37 $m^2 g^{-1}$ and unique mesoporous features, which are beneficial to realize high density of catalytic active sites, and efficient mass transfer and diffusion in ORR catalysis. Due to the nitrogen-containing functional group of lignin, nitrogen doping was verified in the lignin-derived carbon, which can modify the electronic characteristics and modulate the local charge density distribution to realize superior ORR activity. Benefiting from these well-designed structures and compositions, the lignin-derived carbon material exhibited pronounced ORR catalytic activity and tolerance to methanol, thus holding great application potentials as ORR catalysts and catalyst supports.

Acknowledgments

This work was financially supported by the Special Project in the Key Areas of General Universities in Guangdong Province (No.2022ZDZX3069), the Post-doctoral Later-stage Foundation Project of Shenzhen Polytechnic University (No. 6023271007K1), and the Scientific Research Startup Fund for Shenzhen High-Caliber Personnel of SZPU (No. 6023330011K).

References:

[1] Ana Belen Jorge, Rhodri Jervis, Arun Prakash Periasamy, Mo Qiao, Jingyu Feng, Linh Ngoc Tran, Maria-Magdalena Titirici 2020 *Adv. Energy Mater.* **10** 1902494

[2] Fantao Kong, Xiangzhi Cui, Yifan Huang, Heliang Yao, Yafeng Chen, Han Tian, Ge Meng, Chang Chen, Ziwei Chang, Jianlin Shi 2022 *Angew. Chem.* **134** e202116290

[3] Chuangang Hu, Rajib Paul, Quanbin Dai, Liming Dai 2021 *Chem. Soc. Rev.* **50** 11785

[4] Hui Luo, Jesús Barrio, Nixon Sunny, Alain Li, Ludmilla Steier, Nilay Shah, Ifan E. L. Stephens, Maria-Magdalena Titirici 2021 *Adv. Energy Mater.* 2101180

[5] Han Young Jung, Jeong Seok Lee, Hyun Taek Han, Jaehan Jung, KwangSup Eom, Jung Tae Lee 2022 *Polymers* **14** 673.

[6] Long Qie, Wei Min Chen, Zhao Hui Wang, Qing Guo Shao, Xiang Li, Li Xia Yuan, Xian Luo Hu, Wu Xing Zhang, Yun Hui Huang 2012 *Adv. Mater.* **24** 2047

[7] Lanlan Wang, Zhenpeng Liu, Jian Zhang 2022 *Nanoscale* **14** 13473

[8] Chuangang Hu, Rajib Paul, Quanbin Dai, Liming Dai 2021 *Chem. Soc. Rev.* **50** 11785

[9] Qiangmin Yu, Jiaoxing Xu, Chuxin Wu, Lunhui Guan 2015 *RSC Adv.* **5** 65303

[10] Xiaoyu Wu, Songmei Li, Bo Wang, Jianhua Liu, Mei Yu 2017 *Micropor. Mesopor. Mater.* **240** 216

[11]Guiju Tao, Lingxia Zhang, Lisong Chen, Xiangzhi Cui, Zile Hua, Min Wang, Jiacheng Wang, Yu Chen, Jianlin Shi 2015 *Carbon* **86** 108

[12] Dohyeon Yoon, Jieun Hwang, Wonyoung Chang, Jaehoon Kim 2018 *ACS Appl. Mater. Interfaces* **10** 569

Advancements in perovskite solar cell efficiency through blade coating techniques and post-treatment optimization

Ziyue Qiang[1,2,3], Yao Wu[2,3], Caixia Wang[2], Xiang Gao[2], Jihong Li[2], Shuang Sun[2], Ningbo Dong[2], Jun Jiang[1] and Xiaoxia Zhao[2,4]
[1]Department of Chemical Physics, University of Science and Technology of China, Hefei, Anhui 230026, China
[2]State Power Investment Corporation Research Institute, South Park, Beijing Future Science Park, Beijing 102209, China
[3]These authors contributed equally to this work and should be considered co-first authors
[4]Corresponding author: zhaoxiaoxia@spic.com.cn

Abstract. Addition of lead chloride ($PbCl_2$) improves the crystallization kinetics of perovskites by suppressing the formation of δ-phase during large-area perovskite film fabrication. However, this addition results in formation and accumulation of lead iodide (PbI_2) platelet atop the film, which adversely affects carrier transport. To eliminate the negative effect of PbI_2 accumulation and simultaneously obtain a dense, mirror-like perovskite film, a post-treatment method involving blade coating formamidinium iodide (FAI) under air atmosphere is used. Instead of being washed away, as reported in literatures using spin coating process, PbI_2 platelets are converted into cesium lead halide perovskites ($CsPbX_3$) through reaction with Cs and halide present in the original perovskite $Cs_{0.22}FA_{0.78}Pb(I_{0.85}Br_{0.15})_3$. An efficiency enhancement of (>10%) is achieved by FAI treatment, which does not require fast solvent removal or nitrogen (N_2) atmospheres, indicating that FAI blade coating is an effective and straightforward method for the post-treatment of excess PbI_2.

1. Introduction

Hybrid organic–inorganic metal halide perovskite solar cells (PSCs) have attracted considerable attention from academia and industry because of the inherently excellent photoelectric properties of perovskite materials and their substantial potential for reducing manufacturing cost [1,2]. The fabrication of PSCs via low-cost solution processes holds the promise for their subsequent commercialization [3]. So far, considerable efforts have been devoted to developing high-quality and uniform perovskite films on a large scale via scalable solution processes, including blade and slot die coatings. Ink engineering involving solvents and additives plays a pivotal role in film quality and stability [4]. $PbCl_2$ has been widely used as an additive for improving perovskite-film quality through crystallization regulation and defect passivation [5,6]. However, the addition of $PbCl_2$ results in the formation of PbI_2 via halide substitution, which has a controversial effect on device performance [7].

The presence of a moderate amount of PbI_2 in perovskite films is considered to passivate defects at perovskite grain boundaries, contributing to reduced carrier recombination [8]. Conversely, some studies reported that PbI_2 platelets on the surface of a perovskite film impeded charge transport owing to the formation of charge-transport barriers [9]. Accordingly, various strategies have been developed to eliminate surface PbI_2, such as chemical polishing and surface treatment with organic salts. However, most of these studies employed the spin-coating method, which is incompatible with large-scale production [10,11].

This study reported the development of a blade-coating method to mitigate the detrimental effects of excess PbI_2 on PSC performance. This method involves the application of a formamidinium iodide (FAI) solution atop the perovskite film, which does not require rapid solvent removal or the N_2 atmosphere. The applied method facilitates the conversion of PbI_2 platelets into $CsPbX_3$ perovskites through a reaction of PbI_2 with Cs and halide present in the original perovskite $Cs_{0.22}FA_{0.78}Pb(I_{0.85}Br_{0.15})_3$. This post-treatment with FAI considerably enhances carrier extraction, resulting in an efficiency enhancement of >10%.

2. Experimental

2.1. Material

N,N-dimethylformamide (DMF, anhydrous, 99.8%), 1-Methyl-2-pyrrolidinone (NMP, 99.5%), isopropanol (IPA, 99.5%) and lithium fluoride (LiF, 99.99%) were purchased from Sigma-Aldrich. Methylamonium chloride (MACl, 99.5%), Lead Bromide ($PbBr_2$, 99.99%), Cesium iodide (CsI, 99.99%), FAI (99.99%), $PbCl_2$ (99.99%), PbI_2 (99.99%), C_{60} (99.99%), bathocuproine (BCP, 99.99%) and [4-(3,6-dimethyl-9H-carbazol-9-yl)butyl]phosphonic acid (Me-4PACz, 99%) were purchased from Xi'an Yuri Solar Co., Ltd. Without purification, all chemicals were utilized exactly as received, unless otherwise specified.

2.2. Precursor materials preparation

Preparation of perovskite-precursor solutions (1.4 M) was carried out according to the stoichiometric composition of $Cs_{0.22}FA_{0.78}Pb(I_{0.85}Br_{0.15})_3$ perovskite materials with a further incorporation of 3 mol% $MAPbCl_3$. These materials were allowed to dissolve in a mixed solvent comprising DMF and NMP (volume ratio = 93:7). To the prepared perovskite-precursor solutions, 1-5 mol% of $PbCl_2$ was added, and the mixture was subjected to overnight stirring at room temperature. The FAI solution was prepared by dissolving FAI (3 mg) in 1 mL of IPA.

2.3. Solar cells fabrication

ITO substrate preparation: A detergent solution, purified water, and ethyl alcohol solution were utilized to ultrasonically clean the ITO glass for 20 min each. UVO treatment was applied for 20 min after air-blowing drying.

The hole transporting layer and perovskite film deposition: A single coating step was used to cast both the perovskite absorber and the hole-selective contact. A solution of perovskite precursors was supplemented with Me-4PACz at 0.25 mg/mL concentration, which was then completely dissolved. Blade coating was then utilized to deposit the perovskite precursor onto the ITO substrate at a rate of 5 mm/s and a distance of 150 μm between the blade and substrate. N_2 gas was used to blow dry the as-prepared perovskite substrate. The angle between the N_2 knife and substrate was approximately 90°, with a 1 cm distance between the surface of the substrate and the outlet of the knife. With a flow rate of approximately 2 mm/s, the gas pressure was maintained approximately 0.2 MPa. Subsequently, the substrates were annealed using a hotplate at 100 °C for 60 min. For the samples undergoing FAI post-treatment, the FAI solution was blade coated onto the perovskite films using the same setup as the one used in the perovskite coating process, followed by annealing at 100 °C for 10 min. All processes were conducted at room temperature under air ambient with ~20% relative humidity.

The electron-transporting layer and electrode deposition: The deposition of C_{60}, LiF, and BCP layers was performed via thermal evaporation under $<2.0 \times 10^{-5}$ Pa vacuum. The evaporation rate was controlled at 0.1 Å/s to achieve layer thicknesses of 15, 0.8, and 7 nm, respectively. Subsequently, a Ag electrode with 80 nm thickness was thermally evaporated using a patterned mask under a high vacuum ($<2 \times 10^{-5}$ Pa). The evaporation rate was 0.1 Å/s for the first 10 nm and 1 Å/s for the remaining 70 nm.

2.4. Characterizations

The perovskite films were investigated for their surface morphologies and microstructures using field-emission scanning electron microscopy (FESEM, Zeiss Sigma 300). The X-ray diffractometer (XRD, D8 Advance) was also utilized for the characterization of perovskite films. Utilizing a Keithley B2901A source meter, the J-V curves of the PSCs were obtained. A solar simulator (Enlitech, SS-F5-3A) was utilized as the illumination source to match AM 1.5 G. The light intensity was measured and calibrated to 100 mW cm^{-2} by utilizing a standard silicon reference solar cell (SRC-2020-KG3 Enlitech). All the solar cells were examined through a black metal aperture with a 0.16 cm^2 mask area.

3. Results and discussion

To attain dense, mirror-like perovskite films, various amounts of $PbCl_2$ were added to the perovskite-precursor solutions. This method is known to suppress the formation of δ-phase by inducing the formation of an additional PbX_2•0.5NMP•0.5DMF adduct [5]. The SEM images of perovskite films with varying amounts of $PbCl_2$ additives are shown in Figure 1. The formation of PbI_2 platelets due to the addition of $PbCl_2$ was observed, and the PbI_2 platelet amount increased with increased $PbCl_2$ amount. To eliminate the possibility of degradation of perovskite into PbI_2 during the 60-min post-annealing process, perovskite films were prepared under identical conditions but with the hotplate annealing period reduced to 20 min. No considerable difference in PbI_2 platelet density was observed via SEM measurement, indicating that the PbI_2 platelets were formed owing to $PbCl_2$ addition via halide substitution instead of hotplate annealing. With increasing $PbCl_2$ addition, the perovskite films became more homogeneous, and a mirror-like film surface was achieved with addition of 5 mol% of $PbCl_2$, as shown in Figure 1 (d).

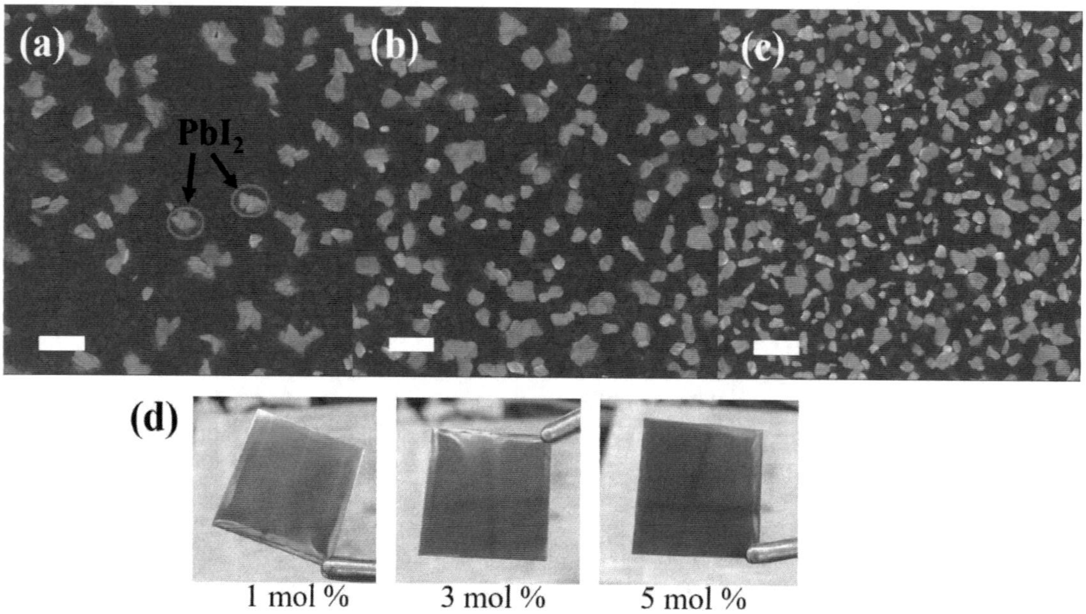

Figure 1. SEM images of perovskite films with additions of (a) 1, (b) 3 and (c) 5 mol% of $PbCl_2$ (scale bar, 1um) and (d) the corresponding photographs. The SEM images was recorded using the InLens detector.

Figure 2 shows the impact of the $PbCl_2$ amount on solar cell characteristics. This figure shows that conversion efficiency is enhanced when $PbCl_2$ addition is increased from 1 mol% to 3 mol%, suggesting that a moderate PbI_2 amount positively affects device performance. However, a substantial decrease in open-circuit voltage as well as short-circuit current is observed as the $PbCl_2$ addition increases from 3 mol% to 5 mol%. This decrease is attributed to the high density of PbI_2 platelets at the top surface of the perovskite films, which adversely affects charge transport and increases series resistance [9].

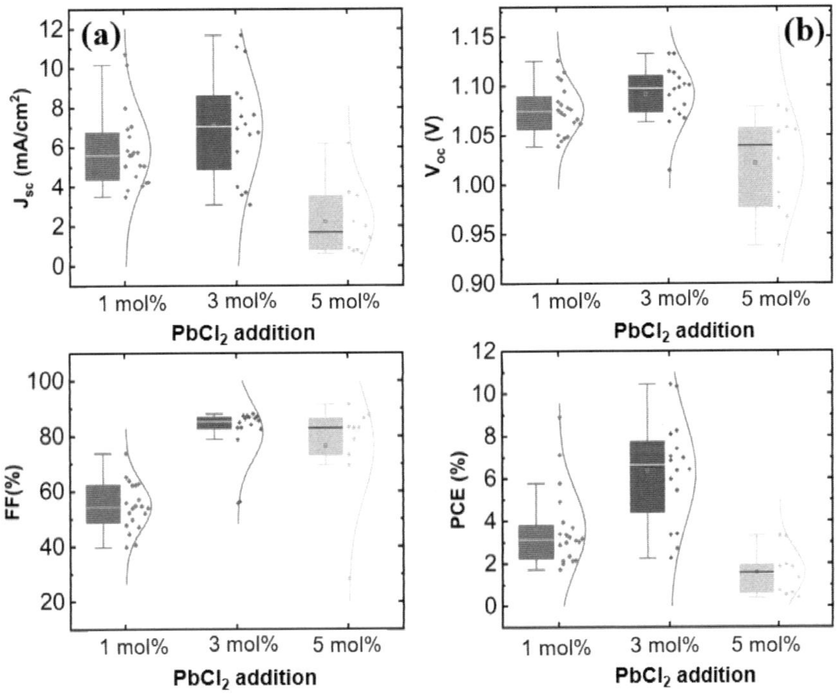

Figure 2. Statistical distribution of the (a) J_{sc}, (b) V_{oc}, (c) FF, and (d) PCE of the solar cells fabricated using perovskite films with varying amounts of $PbCl_2$ addition.

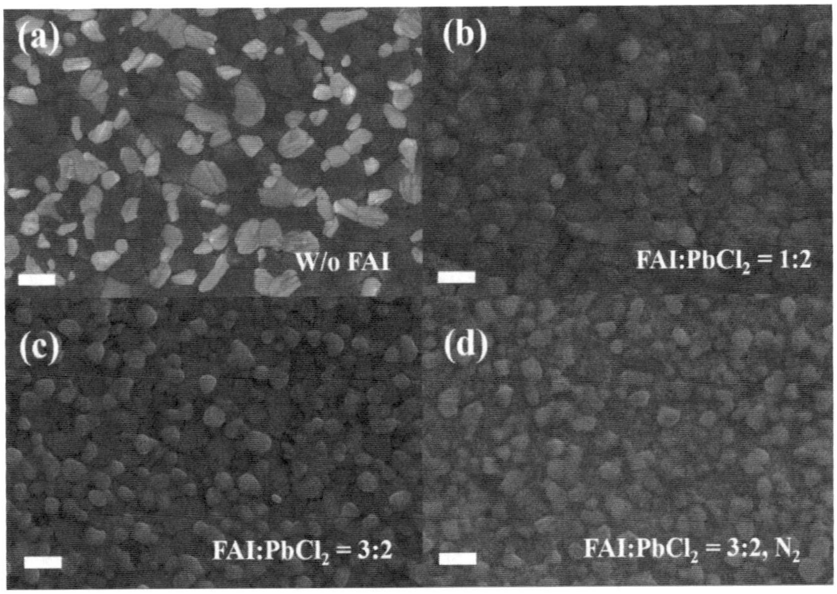

Figure 3. SEM images of perovskite films with addition of (a) 5 mol% of $PbCl_2$ before FAI treatment. Perovskite surface morphology after FAI treatment with (b) FAI:$PbCl_2$ = 1:2 and (c) 3:2 (molar ratio). (d) Morphological characteristics of perovskite films treated by the FAI solution in a N_2 glove box (scale bar, 400 nm).

To address the negative effects of PbI_2 accumulation, an FAI post-treatment was conducted by blade coating a 3-mg/mL FAI solution on perovskite films with $PbCl_2$ (5 mol%) addition. Figure 3 shows that the PbI_2 platelets with high brightness were supplanted by ~200 nm–diameter nanoparticles. The density of these particles was directly proportional to the FAI amount. Moreover, the FAI treatment appeared to widen the gap between perovskite grains, indicating that FAI reacts with PbI_2 platelets as well as with the perovskite. This post-treatment process conducted in a N_2 glove box displayed no considerable difference in perovskite surface morphology compared with that treated under the air condition, as shown in Figure 3 (d).

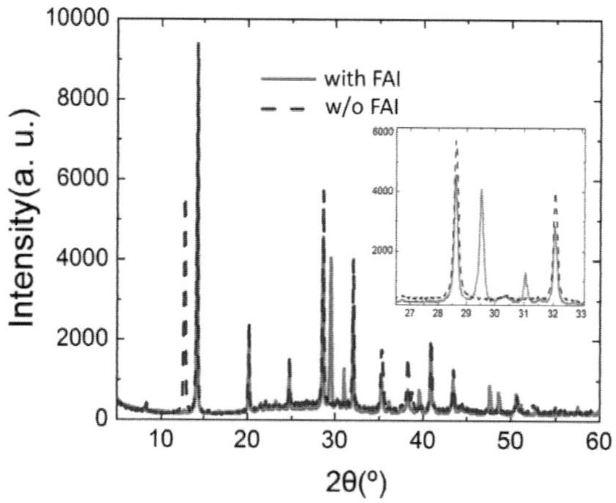

Figure 4. XRD spectra of perovskite films with and without FAI post-treatment.

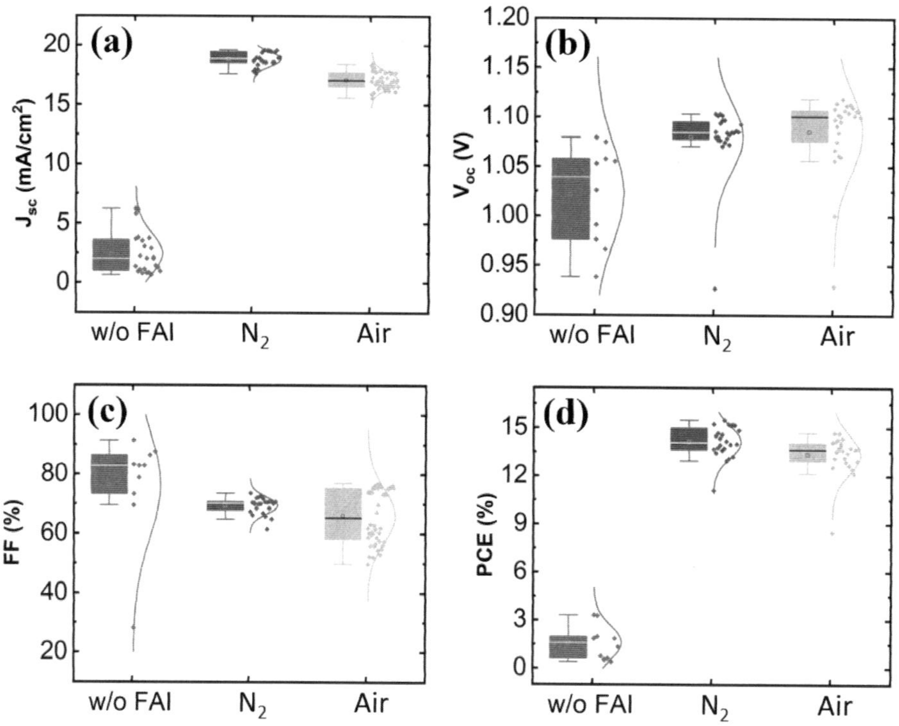

Figure 5. Statistical distribution of the (a) J_{sc}, (b) V_{oc}, (c) FF, and (d) PCE of the solar cells fabricated using perovskite films before and after FAI treatment; N_2 denotes the treatment under N_2 atmosphere in a glove box and air denotes the treatment under air atmosphere.

The X-ray diffraction (XRD) spectra presented in Figure 4 reveal a strong PbI_2 diffraction peak at 12.7° for the perovskite with 5 mol% of $PbCl_2$ addition. For the perovskite treated with FAI under the air atmosphere, this PbI_2 peak is absent, which is consistent with the findings observed through SEM. Two additional XRD peaks are observed at 29.5° and 31.0°, likely attributed to the formation of cesium lead halide ($CsPbX_3$) perovskites as a product of the FAI treatment [12,13].

Figure 5 shows that the open-circuit voltage as well as short-circuit current are substantially improved by FAI post-treatment under N_2 and air atmospheres, resulting in an efficiency enhancement of >10%. The open-circuit voltages of the samples processed under the air atmosphere are marginally higher than those processed in a N_2 glove box. This finding indicates that FAI blade coating is an effective and straightforward method for the post-treatment of excess PbI_2. The device performance could be further enhanced by applying a passivation process during FAI blade coating-a topic that will be explored in future studies.

4. Conclusion

In this work, we developed a blade-coating method to mitigate the detrimental effects of excess PbI_2 on PSC performance. This method involves the application of a FAI solution atop the perovskite film, which does not require rapid solvent removal or N_2 atmosphere. The PbI_2 platelets were converted into $CsPbX_3$ perovskites through a reaction of PbI_2 with Cs and halide present in the original perovskite films. An efficiency enhancement of >10% is achieved after FAI treatment, indicating that FAI blade coating is an effective and straightforward method to mitigate the detrimental effects of excess PbI_2 on PSC performance.

5. Acknowledgments

This work was supported and funded by State Power Investment Corporation Research Institute (042300034956, C-XNY-202203).

6. References

[1] Stranks S D, Eperon G E, Grancini G, Menelaou C, Alcocer M J, Leijtens T, Herz L M, Petrozza A and Snaith H J 2013 *Science* **342** 341-344
[2] Shi D et al 2015 *Science* **347** 519-522
[3] Yang Z et al 2021 *Sci. Adv.* **7**(18) eabg3749
[4] Bowei Li and Wei Zhang 2023 *J. Phys. Energy* **5** 032002
[5] Bu T et al 2021 *Science* **372**(6548) 1327-1332
[6] Kim J E, Kim S S, Zuo C, Gao M, Vak D and Kim D Y 2019 *Adv. Funct. Mater.* **29** 1809194
[7] Gao Y, Raza H, Zhang Z, Chen W, Liu Z 2023 *Adv. Funct. Mater.* **33** 2215171
[8] Chen Q, Zhou H, Song T B, Luo S, Hong Z, Duan H S, Dou L, Liu Y and Yang Y 2014 *Nano Lett.* **14** 4158–63
[9] Xu K et al 2022 *ACS Energy Lett.* **7**(10) 3600-3611
[10] Luo D et al 2018 *Science* **360**(6396) 1442-1446
[11] Zhao L et al 2022 *J. Am. Chem. Soc.* **144**(4) 1700-1708
[12] Yin G, Zhao H, Jiang H, Yuan S, Niu T, Zhao K, Liu Z and Liu S 2018 *Adv. Funct. Mater.* **28**(39) 1803269
[13] Ding W, Liu H, Zhang S, Qiu D, Li X and Wang S 2022 *Adv. Funct. Mater.* **32**(8) 2105164

Effects of strain induced martensite on high temperature electrochemical performance of 321 stainless steel in the simulated PWR first circuit environment

Dongxing Li[1], Han Cao[1,*], Junxuan Gao[1], Quan Zheng[1], Peng Zhang[1], Weihua Zhong[1]

[1] China Institute of Atomic Energy, Beijing 102413, China

*Corresponding authors.

E-mail addresses: zcaohan@foxmail.com (Han Cao)

Abstract. The electrochemical properties of 321 stainless steel with different cold strains in simulated PWR first circuit environment were studied mainly by electrochemical impedance spectroscopy (EIS). The results show that with the increase of strain (0, 20%), the charge transfer resistance increases and the film resistance decreases. By combining the electrochemical experiments and microscopic characterization results of different specimens, including solution annealed, cold worked and hot worked, it is found that the charge transfer resistance increases with the increase of deformation degree, the increase of strain induced martensite content enriches Cr in the oxide layer, but with the decreased material corrosion resistance.

1. Introduction

321 stainless steel (SS) is one kind of austenitic stainless steel, to which Ti is added as stabilizing element. Due to its excellent high temperature mechanical and corrosion resisting properties, and good post-irradiation toughness reported in some literatures, it is often used as construction material of pressurized water reactor (PWR) first circuit [1-6]. In the process of manufacturing, welding and installation, structural parts would undergo cold deformation to different extents. Since austenitic stainless steel is thermodynamically metastable, the face centred cubic (fcc) austenite phase is transformed to body centred cubic (bcc) martensite phase by cold strain, which is called strain induced martensite, sometimes referred to α '-martensite. Studies have shown that cold strain has a significant effect on the thickness and composition of the oxide film formed on the surface of austenitic stainless steel in corrosive environment [7-9]. But the effects of martensite remain unclear[10]. It is generally accepted that the corrosion of stainless steel is an electrochemical process [11-14]. Electrochemical research methods can be regarded as appropriate and practical for investigating corrosion behaviour.
The aim of present work was to explore the influence of strain induced martensite on high temperature electrochemical performance of 321 stainless steel in the simulated PWR first circuit environment and try to explain the mechanism behind it. The high temperature electrochemical behaviours of 321SS with different martensite fraction are studied mainly by electrochemical impedance spectroscopy.

2. Experimental procedures

2.1. Material and specimen

The material used in this study is commercial grade 321SS (06Cr18Ni11Ti). The chemical composition is shown in Table 1. The parent metal was supplied as a column with a diameter of 80mm, the as-received microstructure was solid solution, remarked as SS. Pieces with a geometry of $10\times60\times40mm^3$ were cut off from the column by wire-electrode cutting. One piece was cold rolled at room temperature to achieve a reduction of 20% in thickness, remarked as CW20. Another piece was also rolled to achieve 20% thickness reduction but under 180℃, remarked as HW20. Specimens with geometry of $6\times6\times0.6mm^3$ were cut from the central part, to avoid the inhomogeneity of microstructure which may be introduced by cold work and hot work and make the specimens representative for the average property of the corresponding material status.

After sampling, both sides of the specimens were sequentially ground up to 2000# with SiC paper. Then were mechanically polished with diamond agent of different particle diameters to achieve mirror-like surface. The scanning electron microscope (SEM) photos of different samples' surfaces are shown in Figure 1. Tensile specimens were also taken from the materials of different status to measure the mechanical properties.

Table 1. Compositions (wt%) of 321SS material.

C	Mn	P	S	Si	Ni	Cr	Ti	Mo	Fe
0.041	0.760	0.030	0.001	0.510	9.010	17.200	0.200	-	Bal.

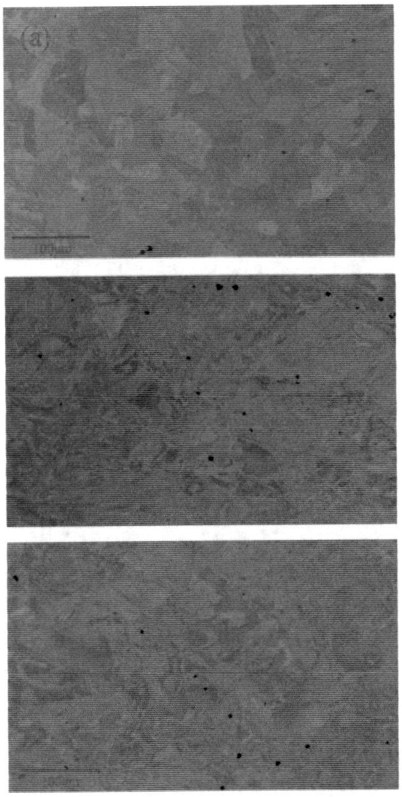

Figure 1. SEM photos of different specimen after polishing. (a) SS (b) CW20 (c) HW20

Journal of Physics: Conference Series **2749** (2024) 012005 doi:10.1088/1742-6596/2749/1/012005

2.2. Phase content measurement and mechanical property characterization

X-Ray Diffraction (XRD) was used to measure the weight fraction of different phases (fcc austenite phase and bcc martensite phase). A light spot of $1\times5mm^2$ was placed in the central part of the specimen to serve as the observation field. Compared to transmission electron microscope (TEM), XRD can measure the phase content in a macro large zone which makes the results statistically representative.

The tensile tests were carried out on a Zwick tensile testing machine at room temperature in atmosphere. The specimens were deformed at strain rate of $2.5\times10^{-4}s^{-1}$. The machine is equipped with laser extensometer to measure the sample strain accurately. The mechanical properties of different samples, like tensile strength and offset yield strength, can be obtained directly or calculated from the stress-strain curve.

2.3. High temperature electrochemical experiments

High temperature electrochemical experiments were carried out in an autoclave with a volume of 2.0L made of 625 alloy. The autoclave is equipped with a recirculating water chemistry loop which refresh the solution in the autoclave at a flow rate of approximately 2 volume per hour. The temperature and pressure in the autoclave and the water chemistry parameters of inlet and outlet water, including dissolved oxygen (DO), pH and conductivity, were monitored continuously. DO was controlled by continuously bubbling nitrogen into the water tank of the loop. The environment parameters are shown in Table 2.

Table 2. Environment of high temperature electrochemical testing.

Parameters	Value
Temperature	330℃
Pressure	15.5MPa
B	1200ppm
Li	2.2ppm
DO	<10ppb

The electrochemical behaviour was tested using a three-electrode system. A $1\times1\times0.2mm^3$ platinum sheet was used as counter electrode and the test sample was regarded as working electrode. Both the platinum sheet and the specimen were welded to nickel wires which were covered by zirconium oxide ceramic that helped insulate the wires from the solution and the autoclave. The wires run though the bottom of the autoclave and were connected to the electrochemical workstation together with the wire of the reference electrode. The latter is $Cu/Cu_2O/ZrO_2$ type, and its potential vs. the SHE potential is about -299mV at the testing temperature.

After specimen installation, the autoclave was closed and filled up with solution described above. When it was pressurized and heated to the target pressure and temperature, the testing system was stabilized for 20 hours. Then electrochemical impedance spectroscopy (EIS) was performed. Finally, the specimen was polarized.

3. Results and discussion

3.1 Phase content and mechanical properties

XRD quantitative phase analysis was completed by adiabatic method. In a multiphase system with N phases and no amorphous phase, if all the phases are known and the K value (RIR) of each phase is known, the diffraction intensity of each phase is measured, and one of the phases is selected as the internal phase, N-1 equations can be listed as following:

$$\frac{I_j}{I_i} = K_i^j \frac{\omega_j}{\omega_i} \text{ (j=1, 2,}\cdots\text{, N-1)} \quad (1)$$

As$\sum_j^n \omega_j = 1$, the mass fraction of each phase can be written as:

$$\omega_j = \frac{I_j}{K_i^j \sum_{i=1}^{N} \frac{I_j}{K_i^j}} \tag{2}$$

The intensity of the diffraction peak can be obtained directly from the diffraction pattern, and K value can be referred to PDF-card. The mass fraction of austenite phase and martensite phase of different specimens and corresponding mechanical properties are shown in Table 3.

The microstructure in the parent metal is at solid solution state, so only a small amount of martensite can be detected in SS. α '-martensite is introduced to the matrix by cold strain. The transformation from austenite to martensite is supressed by heating, so the martensite fraction in CW20 is higher than HW20.

After cold work and hot work, the grains become shattered and twisted, as shown in Figure 1. Due to the work hardening effect, the tensile strength (σ) and offset yield strength ($\sigma_{0.2}$) rises in CW20. And the work hardening effect is weakened when the sample is heated when deformation occurs. So the σ and $\sigma_{0.2}$ of HW20 are lower than CW20 but share not much difference. That's because the relatively low heating temperature.

The transition process from austenitic to martensite can be supressed at rather low temperatures, but some processing procedures are usually carried out at room temperature for economic reasons or to make use of work hardening effect to achieve synthetical mechanical properties. The phase transformation seems inevitable. And once martensite is introduced to the matrix, it can be only totally removed after a period of heat preservation at 1100℃, which is far above the working temperature of 321SS in a nuclear reactor[15]. So, the potential impacts of martensite still exist.

Table 3. α' martensite fraction and tensile properties of different specimens.

	α' martensite fraction (×100%)	$\sigma_{0.2}$ (MPa)	σ (MPa)
SS	/	226	592
CW20	0.22	657	768
HW20	0.11	540	690

3.2 High temperature electrochemical experiment

EIS spectra is performed to describe the corrosion behaviour. The Nyquist diagrams of SS, CW20, HW20 are shown in Figure 2. The Nyquist plots of three samples are consisted of two parts, in which the high frequency region corresponding to the left of the graph presents a semicircle, while the low frequency region corresponding to the right part shows an incomplete semicircle. Since the shape of the curve is semicircle both in high frequency region and low frequency region, the corrosion behaviour is controlled by charge transfer process.

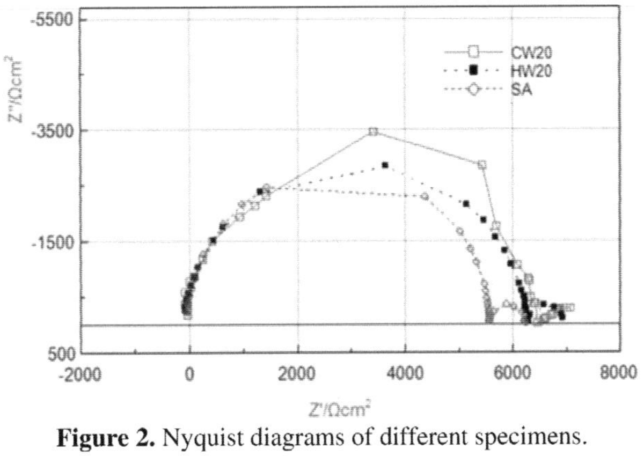

Figure 2. Nyquist diagrams of different specimens.

The equivalent electro circuit (EEC) used to fit EIS is shown in Figure 3. The R_S is the buffer solution resistance. The CPE is the constant phase angle element replacing ideal capacitance due to non-uniform distribution of current and potential caused by non-uniformity, roughness of the electrode surface and other factors. R_{ct} is used to represent charge transfer resistance and R_f for passivating film resistance on the specimen surface due to the film resistance model. The values of each EEC element of different samples are listed in Table 4.

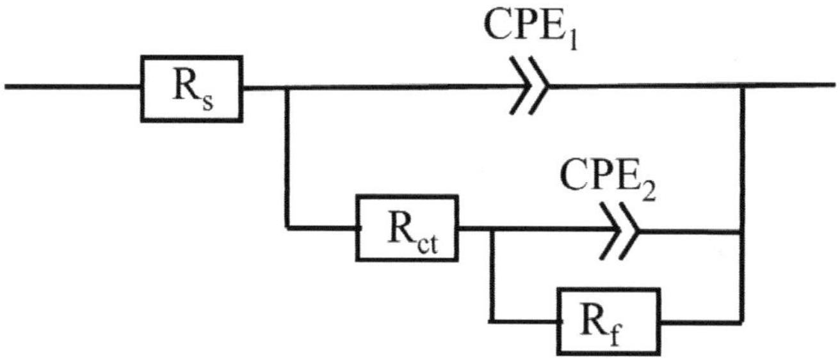

Figure 3. EEC model for EIS spectra.

Table 4. EEC elements' values of each specimen.

	Rs (Ω/cm²)	T_1(×10^{-9})	P_1 (×1)	R_{ct} (Ω/cm²)	T_2(×10^{-4})	P_2 (×1)	R_f (Ω/cm²)
SS	15.5	8.26	1.0	5604.6	7.60	1.0	659.8
HW20	20.6	14.9	0.93	6392.2	33.70	1.0	647.7
CW20	19.8	6.04	1.0	6414.9	27.91	1.0	622.7

Similar values of R_{ct} are shared by CW20 and HW20, larger than that of SS. The grains are deformed, shattered, twisted and fiberized under strain, and dislocations and residual stress are introduced to the matrix. The transfer of charge from base metal to the solution is hindered by these changes in microstructure. Since the work hardening effect shares almost the same mechanism, CW20 and HW20 with proximal σ and $\sigma_{0.2}$, having similar R_{ct} is reasonable.

With the mass fraction of martensite rising, the R_f drops, which indicates a decrease in corrosion resistance of the passivating film. From Figure 4 which shows the surfaces of different samples after polarization, specimens after strain are more severely corroded, but the corrosiveness of CW20 and HW20 cannot be differed from the morphology.

3.3 *Surface and cross section elements distribution analysis of different samples*

After high temperature electrochemical test, the element distribution on the surface and cross section is analysed by energy dispersive spectrum (EDS). From the surface EDS results of element mass content shown in Table 5, it can be seen the Cr content in the film surface increases with martensite fraction riseing. The SEM photo and EDS maps of different elements on the surface of CW20 are shown in Figure 5. C and O are concentrated at the oxide particles. The concentration of Fe in the oxide particles is higher than that in the oxide film. In contrast, Cr and Ni concentrations are lower in these particles.

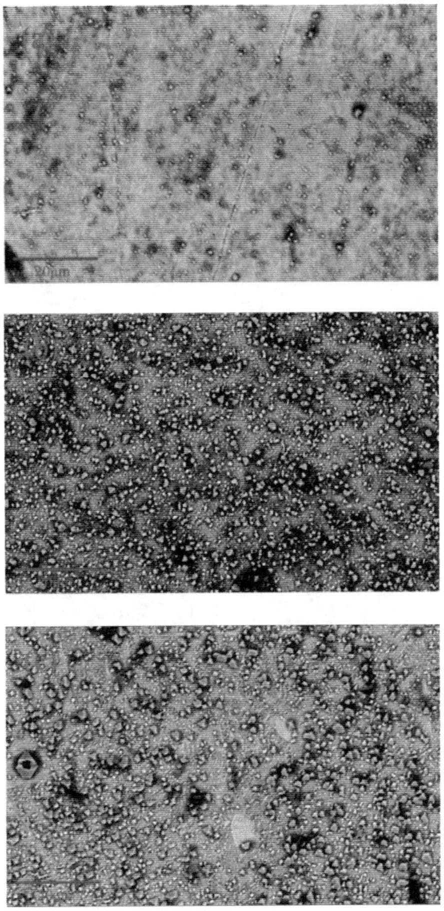

Figure 4. SEM photos of different specimen after polarization. (a) SS (b) CW20 (c) HW20

Table 5. EDS result of weight content of different elements on the surface.

	O (%)	Cr (%)	Ni (%)	Fe (%)
SS	5.41	18.12	10.21	64.06
HW20	15.82	18.32	7.31	57.96
CW20	15.78	18.38	7.19	57.84

The SEM photo and distribution of Cr on the cross section of three samples are shown in Figure 6. From left to right, is the thermosetting resin, oxide layer and the matrix in turn. It can be seen, in the oxide layer of SS, there is chromium depletion while the oxide layer of HW20 andCW20 are rich in chromium, which is consistent with the surface EDS results. And there is Cr depletion at the interface between the oxide layer and the interface, which can be inferred from a dark stripe between two bright zones.

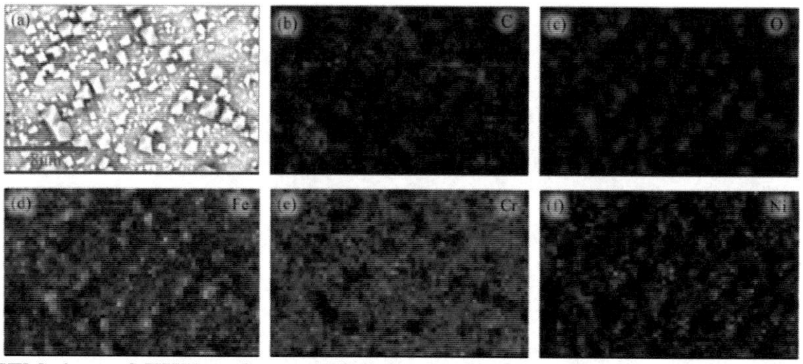

Figure 5. SEM photo of CW20 surface and EDS maps. (a) Surface morphology (b) C (c) O (d) Fe (e) Cr (f) Ni

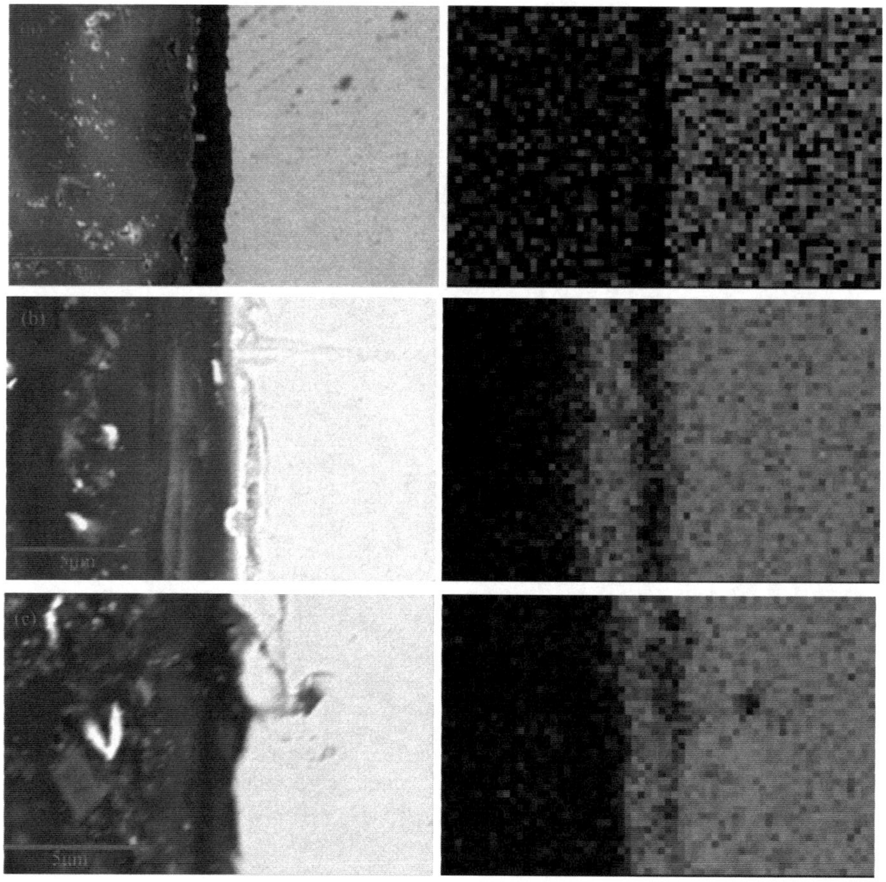

Figure 6. SEM photos and Cr distribution on the cross section of different samples.
(a) SS (b) CW20 (c) HW20

The EDS line scanning of HW20 suggests weight contents of different elements along the cross section, shown in Figure 7. The highest Cr content appears near the outer surface of the oxide layer, and slowly decreases along the thickness. The lowest Cr content appears near the interface between the oxide layer and the matrix, corresponding to the dark stripe, with an O peak near it, indicating that the dark stripe is heavily corroded. The existence of martensite causes a chromium-poor zone at the bottom of the oxide layer, leading to weakened protective effect of passivation film. This phenomenon may be connected with the diffusion coefficient of Cr. It has been reported that the diffusion coefficient of Cr in martensite is higher than that in austenite [16]. Once the oxide film is formed on the surface, Cr more easily diffuses to the film in the 321SS with higher martensite fraction, leaving Cr depletion near the interface and destroying protective effect of the passivation film. In order to verify this conjecture, a more detailed analysis of the oxide film structure and a precise calculation of the diffusion process of Cr are needed.

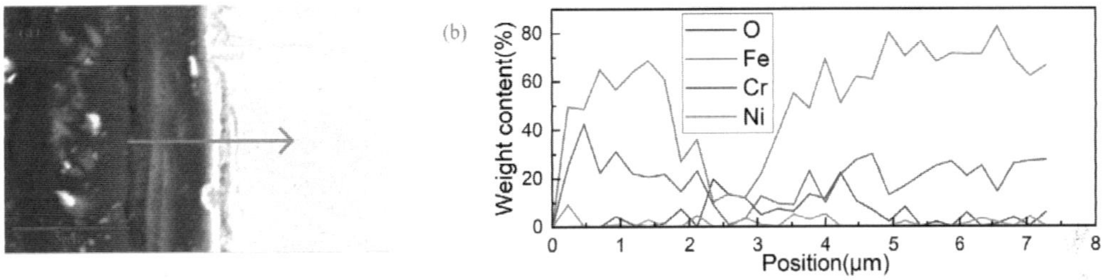

Figure 7. EDS line scanning on the cross section of HW20. (a) Cross section morphology and the line scanning position (b) EDS line profile across the oxide layer

4. Conclusions

In the present study, 321SS with different martensite phase mass fraction are acquired by cold work and hard work. Mechanical properties are tested by tensile testing and electrochemical behaviour in simulated PWR first circuit environment is detected by EIS. Conclusions can be drawn as following:

(1) Transformation from austenite phase to martensite phase is introduced to the matrix by cold strain and is supressed by hot strain.
(2) The electrochemical corrosion behaviour is controlled by charge transfer process.
(3) The charge transfer increases with the increase of strain. The film resistance decreases with the increase of martensite fraction, and the passive layer shows lower protective property.
(4) The existence of martensite creates a chromium-poor zone near the interface between the oxide layer and the interface, and weakens the protective effect of the passivation film.

Acknowledgements

The author would like to acknowledge Chinese Materials Research Scoiety (MRS) for the organization to make the participation in the 1st World Energy Materials Conference feasible. Thanks also to Han Cao and Weihua Zhong for the many comments regarding the manuscript. The author would hereby express his greatest thanks to the Project - Study on IASCC behaviour and prediction model of fastening bolts of reactor components (2019YFB1900902) for the financial support of the presented research.

References

[1] World Nuclear Association.World nuclear power reactors & uranium requirements [EB/OL]. http://www.world-nuclear.org/information-library/facts-and-figures/worlf-nuclear-power-reactors-and-uranium-requireme.aspx.
[2] Chabaud-Reytier M, Allais L, Caes C, Dubuisson P, Pineau A. Mechanisms of stress relief cracking in titanium stabilised austenitic stainless steel. J Nucl Mater 2003;323:123.

[3] A. Abbas, E. Fazakas, T. Torok. Corrosion studies of steel rebar samples in neutral sodium chloride solution also in the presence of a bio-based (green) inhibitor. Int. J. Corros. Scale Inhib 7(2018)38–47.

[4] Han Cao, Junxuan Gao, Wanhuan Yang, Cong Liu, Dongxing Li, Peng Zhang, Quan Zheng, Weihua Zhong. Effect of heavy ion irradiation on the electrochemical behavior of 321 stainless steel. J Nucl Engineering and Design, 419(2024)112972.

[5] Parvathavarthini N, Dayal RK, Gnanamoorthy JB. Influence of prior deformation on the sensitization of AISI Type 316LN stainless steel. J Nucl Mater 1994;208:251.

[6] Pratik P. Joshi, Nilesh Kumar, Korukonda L. Murty. Materials for Nuclear Reactors. Encyclopedia of Materials: Metals and Alloys, (2022)364-376.

[7] P.L. Andresen, P.W. Emigh, M.M. Morra. Effects of PWR primary water chemistry and deaerated water on SCC, Corrosion, 4 (2005) 1-20.

[8] P.L. Andresen, L.M. Young, Characterization of the Roles of Electrochemistry, Convection and Crack Chemistry in Stress Corrosion Cracking, Proc. 7th International Symposium on Environmental Degradation of Materials in Nuclear Power Systems-Water Reactors, NACE, (1995) 579–596.

[9] P.L. Andresen, P.W. Emigh, M.M. Morra. Effects of PWR primary water chemistry and deaerated water on SCC, Corrosion, 4 (2005) 1-20.

[10] S.K. Pradhan, P. Bhuyan, L.R. Bairi, S. Mandal. Comprehending the role of individual microstructural features on electrochemical response and passive film behaviour in type 304 austenitic stainless steel, J Corros. Sci., 180(2021)109187.

[11] Z. Lu, T. Shoji, Y. Takeda, Effects of loading mode and water chemistry on stress corrosion crack growth behavior of 316L HAZ and weld metal materials in high temperature pure water, Corros. Sci., 50 (2008) 625-638.

[12] H. Yamashita, S. Ooki, Y. Tanaka. SCC growth behavior of BWR core shroud materials, Int. J. Pres. Ves. Pip., 85 (2008) 582-592.

[13] Q.A. Luis, D.-A. Yureel, Avelino-Jiménez, et al. Analysis of the Physicochemical, Mechanical, and Electrochemical Parameters and Their Impact on the Internal and External SCC of Carbon Steel Pipelines. J Mater., 13(24)(2024).

[14] C. García, F. Martín, P.D. Tiedra, Effects of prior cold work and sensitization heat treatment on chloride stress corrosion cracking in type 304 stainless steels, Corros. Sci., 43 (2001) 1519-1539.

[15] Heat Treatment Austenitic Corrosion-Resistant Steel Parts. Society of Automotive Engineers, USA.AMS 2759/4B. 2011, 11.

[16] L. Hongyun, Z. Yubo, L. Hongdou, L. Jinlong, M. Yue, Characterization of the oxide films formed on low temperature sensitized AISI 321 stainless steel with different strain levels in elevated temperature borate buffer solution, J Alloys and Compounds (2017), doi: 10.1016/ j.jallcom.2016.12.107.

Gradient microstructure in Ni-Cr-Mo ultra-heavy steel plate after tempering

L Meng [1], N Zhang [1*], T Pan [2*], B Zhang [1], X B Luo [2], F Chai [2] and Y Tu [1]
[1] Metallurgical Technology Institute, Central Iron and Steel Research Institute, Beijing, 100081, China;
[2] Research Institute of Engineering Steels, Central Iron and Steel Research Institute, Beijing, 100081, China.
*Corresponding author: zhangn@126.com, taopan@vip.sina.com

Abstract. This paper studied through-thickness gradient microstructure of an 80mm ultra-heavy steel plate by multi-dimensional characterizations, meanwhile the cause of microstructure variability as well as its effect on strength was analyzed. Results show major lath-structure in the whole steel plate, while lath morphology and grain boundary distribution are different along the thickness direction. During phase transformation, the higher cooling rate in the surface enhances the forming ability of V1/V4 variant pairs and corresponding 5°-10° boundaries, meanwhile the higher cooling rate and smaller original austenite grain contribute to more nucleation sites and stronger strain incompatibility, which is corresponding to more low-angle boundaries, so the surface layer shows higher low-angle boundaries; Compared with the surface layer, the moderate cooling rate is beneficial for V1/V2 variant pairs, and Σ3 boundaries between the pairs are obtained, leading to higher high-angle boundary density with decreased low-angle grain boundary density. This boundary characteristic corresponds to the thinner and longer lath morphology. With respect to the relationship between mechanical properties and microstructure, more low-angle boundaries in the surface layer and more high-angle boundaries in inner layers contribute to stronger dislocation strengthening and grain boundary strengthening respectively, and realizes the coordinated through-thickness property.

1. Introduction

With the development of marine industry, there is increasing requirement of high performance offshore engineering heavy steel with high strength and toughness. In general, the strength of ultra-heavy steel is improved by Martensite / Bainite phase transformation during quenching, and then the toughness is optimized by tempering treatment [1-3]. However, with the increasing thickness of heavy steel, the deformation and heat transfer non-uniformity along thickness direction may lead to inhomogeneous microstructure at different thickness and poor stability [4-5].

When phase transformation occurs in original austenite grains, lath microstructure may be obtained inside a prior austenite grain and the grain could be divided into several levels: laths with the same habit plane belong to one packet, inside which the group of laths with same orientation forms a block [6-9]. Compared with granular bainite, which is usually believed as a brittle zone to deteriorate toughness, the laths are beneficial for strengthening and toughening. It was reported that the width of block is identified as effective grain size unit to determine the strength [10-14], and the blocks can be characterized by EBSD [15]. Besides block width or effective grain size, grain boundary also plays an important role on mechanical property. Due to the orientation relationship between the laths and prior austenite grains, such as K-S relationship and N-W relationship, specific boundaries could be obtained. A high density of twin-related V1/V2 variant pairs were reported in the surface of high strength ultra-heavily plate steel, so Σ3 (<111>60°) inter-variant boundaries were observed [5]. The adjacent coalesced bainite plates in quenched and tempered low-carbon steels are shown to be V1 and V4 variants, with a misorientation of approximately 10° [16]. In general, high-angle boundaries could inhibit the crack propagation, and the block or sub-block boundaries could hinder the dislocation movement [13,17-18], meanwhile the effect of low-angle boundaries on strength improvement could not be ignored [14,17-18].

For ultra-heavy steel, it is essential to clarify the uneven microstructure distribution along the thickness direction, which may deteriorate the mechanical property uniformity. In this study, gradient microstructure of Ni-Cr-Mo ultra-heavy steel is investigated, and the difference between surface and inner layer is paid special attention. The aim is to elucidate the reason for gradient microstructure and the relationship between microstructure and mechanical properties.

2. Experimental methods

The investigated material is an 80mm industrial tempered low-carbon Ni-Cr-Mo ultra-heavy steel. The main chemical composition of the material is shown in Table 1, and the rolling and heat treatment were carried out as follows: the steels were heated at 1200°C and then hot rolled; afterwards the heat treatment consists of an austenitization at a temperature of 890°C followed by roller-quenching using an industrial roller at room temperature and then tempering at a temperature of 620°C. In order to assess the properties and microstructures of the steel comprehensively, the ultra-heavy steel was cut at different thickness locations, as shown in Table 1. Considering that there may exist obvious circumstance deviation in surface region, such as the occurrence of dynamic recrystallization or decarburization during deformation and heat treatment, this study discussed the microstructure characteristic of near-surface region.

Table 1 The investigated samples.

Samples	Near-surface layer			1/4 layer		Central layer	
Composition (wt.%)	C	Si	Mn	Ni	Cr	Mo	Fe & other elements
	≤0.11	0.10-0.37	0.40-0.70	4.40-5.30	0.40-0.70	0.30-0.67	Balance
80mm steel	A1 (~ 9mm from the surface)			A2 (~20mm from the surface)		A3 (~40mm from the surface)	

The samples were polished and etched, then the microstructures were measured using the GeminiSEM 500 field emission scanning electron microscope and the HKL Nordlys Max3 EBSD detector. The working voltage of the scanning electron microscope is 20kV, the sample inclination angle is 70°, the working distance is 14~18mm, and the EBSD test step length is 0.2-0.6μm. Following M^x microstructure (such as Misorientation, Multiscale, Micro-texture, Multiphase, Multi-levelled, etc.) analysis method, the microstructures were analyzed in terms of morphology, equivalent grain size and shape, grain boundary character distribution, special CSL boundaries and prior austenite grain reconstruction. The correlations between grain size, grain boundary character distribution, strength and toughness were studied in detail. To avoid statistic error, more than 10 areas were measured in each sample, and the calibration rate of EBSD data is higher than 85%.

3. Results and discussions

Figure 1 displays typical microstructures of all samples. Lath-structures are dominant in every sample, meanwhile other types of structure can also be observed, such as granular bainite. This indicates uneven phase transformation during quenching. The lath shape and size differ at different layers, and long and parallel laths are shown at inner layers, including A2 and A3 samples. It needs to be noted that despite of microstructure difference, the fraction of lath-structure is high in every sample. Therefore, this study focuses on grain boundary and grain size, and analyzes the correlation between them and influencing factors as well as mechanical properties.

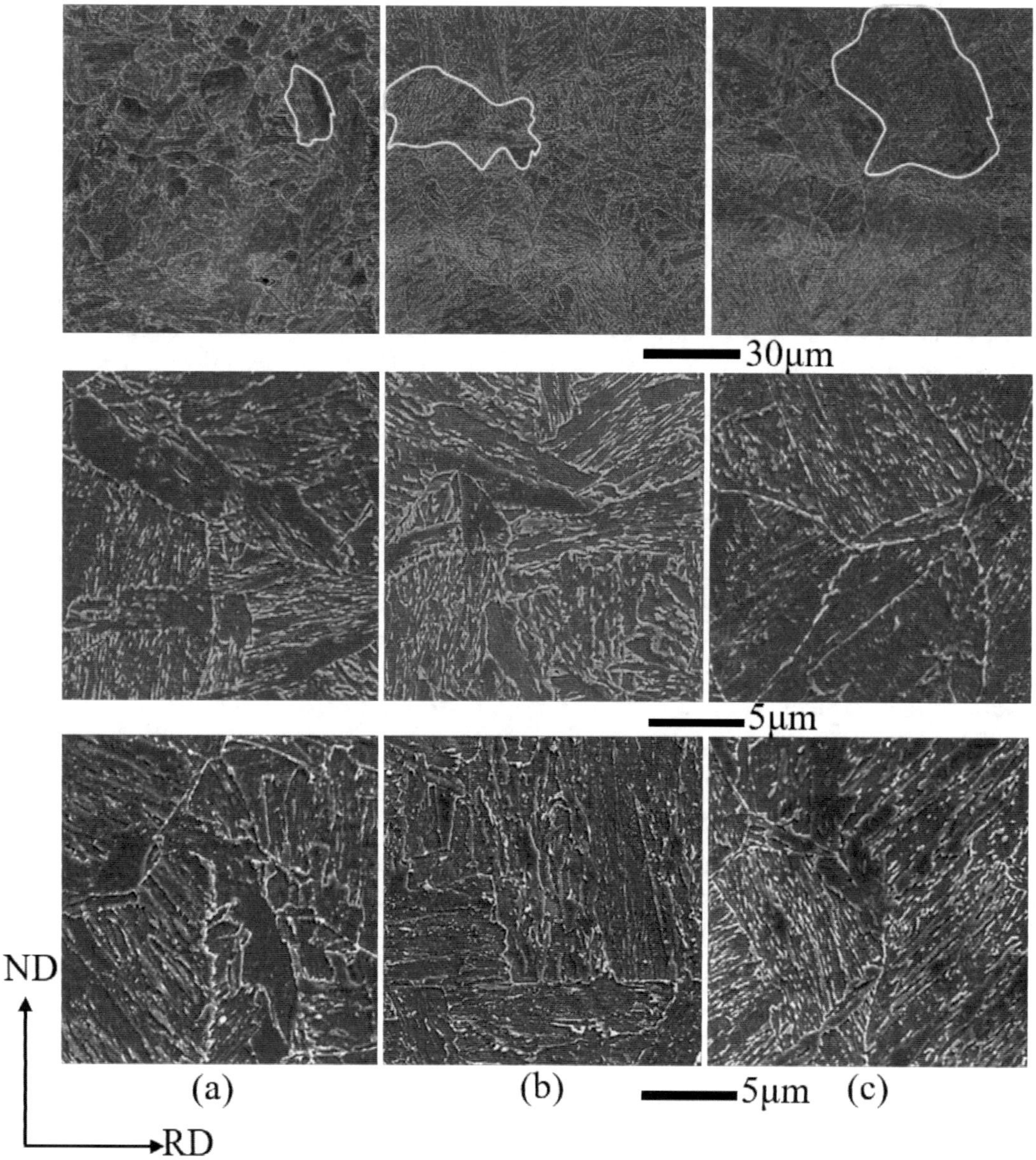

Figure 1 Typical microstructures at different magnification (yellow lines indicate typical prior austenite grain boundaries). (a) A1, (b) A2, (c) A3.

Figure 2 shows orientation maps of all samples, and grain boundaries (misorientation angles ≥15°) are given. Similar to the results in figure 1, much thinner and longer lath-structures are shown in A2 and A3. Using the 'grain size' module of Aztec Crystal processing software, equivalent grain width and shape are calculated by oval fitting, and the results are given in figure 2(d-e). The grain length is the fitted ellipse major diameter, and the grain width is the fitted ellipse minor diameter, then the ratio of grain length/grain width is the rate of these two values. Given that both block and grain are defined by 15° boundaries, the calculated grain width is equivalent to the block width. The fraction of those grains thinner than 3μm is lower in A1, so the average grain width is higher. Correspondingly, the ratio of grain length/grain width is lower in A1, as shown in figure 2(d-e). The grain size and shape in A2 and A3 are close, that is, these values are similar in the inner layers of ultra-heavy steel.

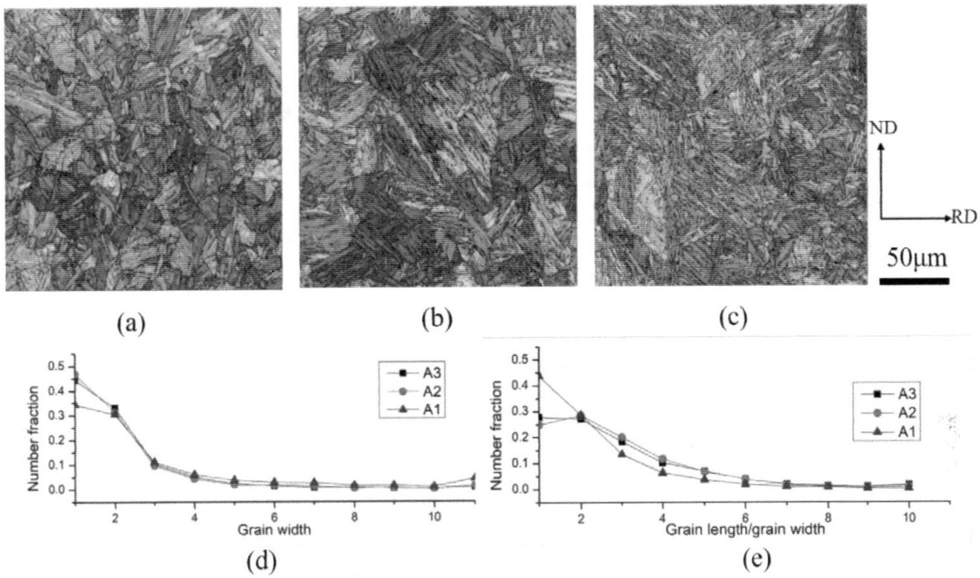

Figure 2 EBSD orientation mappings. (a) A1, (b) A2,(c) A3, (d) equivalent grain width and (e) grain shape characteristics.

The grain size and shape information obtained by EBSD data is in intrinsic correspondence with the high-angle (15°) grain boundaries distribution. Grain boundary distribution may be connected with the strength and toughness, meanwhile it could be effective in analyzing the variant selection behavior during phase transformation. By comparing grain boundaries with different misorientations in different regions in figure 3, it is shown that more low-angle (2°-15°) boundaries are shown in near surface layer, while more >45° boundaries are observed in inner layers, including A2 and A3 samples. >45° high- angle boundaries are believed to be beneficial for the toughness improvement. The fraction of 15°-45° boundaries are similar and not high in different samples.

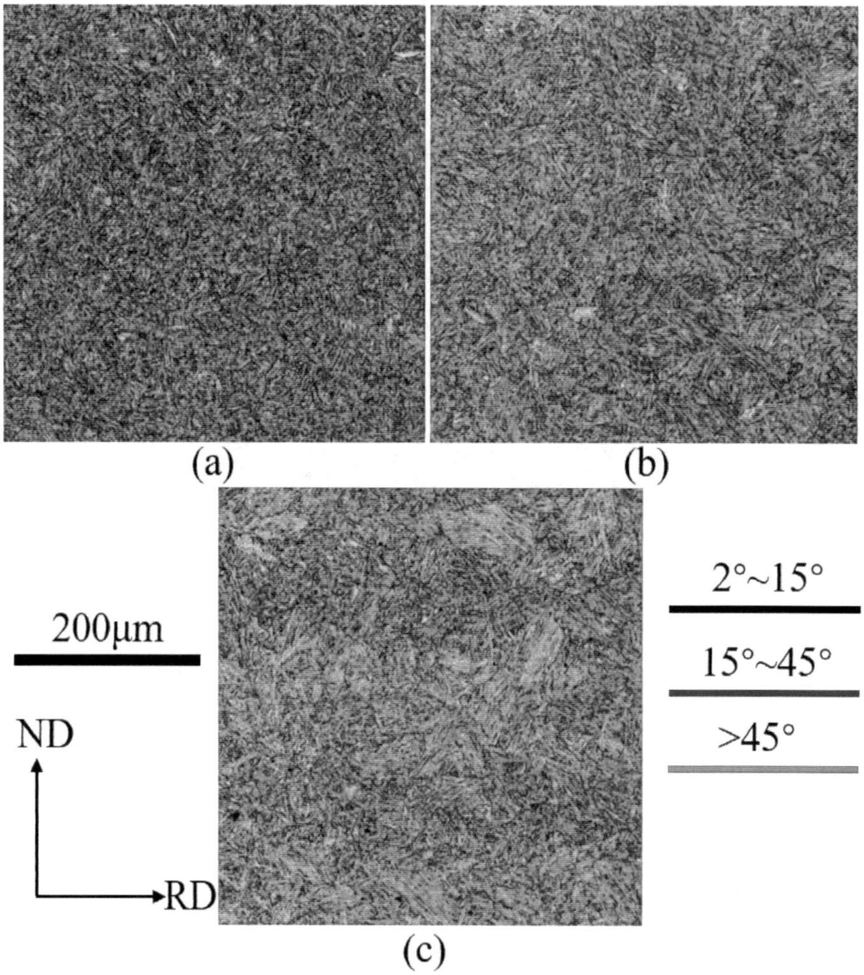

Figure 3 Grain boundary character distributions. (a) A1, (b) A2, (c) A3.

The grain boundary differences may be linked to different variant pairs distribution. The ultra-heavy steel in this study contains about 0.08-0.09wt.% carbon, and K-S relationship is complied. As shown in figure 4, it is obvious that the density of Σ3 grain boundaries in A1 is much lower than those in A2 and A3, and this contributes to the lower >45° grain boundaries in figure 3(a). The occurrence of Σ3 grain boundaries is suggested to be linked with V1/V2 variant pairs, and it could modify the strain during phase transformation. The ability of Σ3 grain boundaries occurrence is determined by phase transformation driving force [19-20]. In this study, the lower fraction of Σ3 grain boundaries near surface may be induced by high cooling rate during quenching. Under this circumstance, much high driving force may lead to rapid phase transformation, while there is not enough time for forming specific V1/V2 variant pairs to relieve the strain [3]. In comparison, the cooling rate in the inner layers of heavy steel is lower, and the strain relieving effect of specific V1/V2 variant pairs could be realized more sufficiently. More V1/V4 variant pairs with the relationship of 7°<011> can be achieved near surface, under higher cooling rate [3, 21].

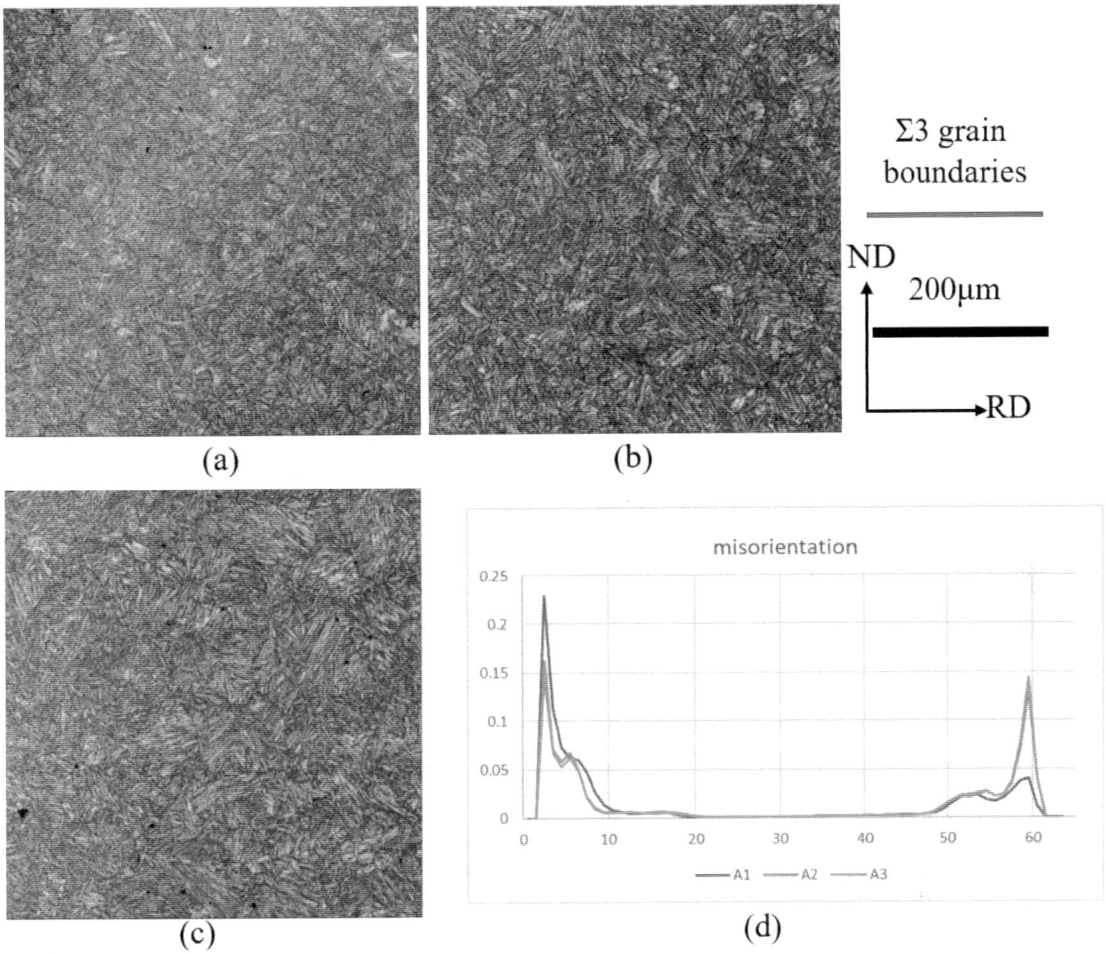

Figure 4 Σ3 boundary distribution (red lines in color pictures). (a) A1, (b) A2, (c) A3, (d) misorientation angle distribution in A1-A3.

The gradient microstructure along the thickness direction is connected to the different variant selection behaviors during phase transformation. Besides the effect of cooling rate at different thickness, the variant selection may also be influenced by prior austenite grains' size and orientation [22]. Figure 5 shows the reconfigured prior austenite grains based on K-S relationship. The texture is calculated, while no strong texture is shown. Regarding the grain size, prior austenite grains are smallest in A1, and a number of coarse prior austenite grains can be observed in A2 and A3. The results are in accordance with that shown in figure 1. During rolling deformation, the strain penetration is reduced from the surface to the center, leading to the coarse microstructure in the inner layers. Small prior austenite grain is not beneficial for obtaining typical lath structure after phase transformation, so the variant quantity may be decreased and Σ3 grain boundaries will be reduced [23].

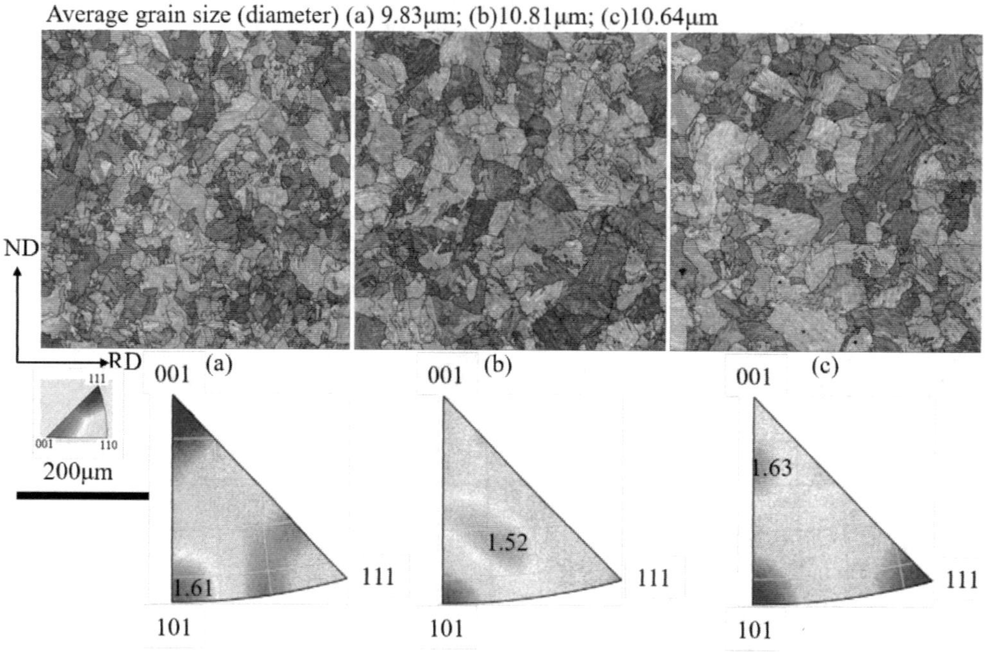

Figure 5 EBSD maps and inverse pole figures of reconstructed prior austenite grains. (a) A1, (b) A2, (c) A3.

The grain boundary distributions in prior austenite grains and tempered samples are counted and shown in figure 6. The higher grain boundary (>15°) fraction in A1 corresponds to the smaller prior austenite grain characteristic, and 2°-5° and 5°-10° grain boundaries are relatively fewer. As discussed earlier, 5°-10° and >45° boundaries are linked to V1/V4 and V1/V2 variant pairs respectively, so the values increase after quenching and tempering. 10°-15° and 15°-45° boundaries are relatively low in quantity and change little after phase transformation. What needs to be noticed is the marked increase of 2°-5° boundaries, especially in A1. 2°-5° misorientation is not related to specific variant pairs, while it is assumed to be linked to phase transformation strain. Both the higher cooling rate and smaller prior austenite grain near surface could impede the phase transformation process and increase nucleation sites, and these are assumed to increase the transformation strain and the 2°-5° boundaries accordingly.

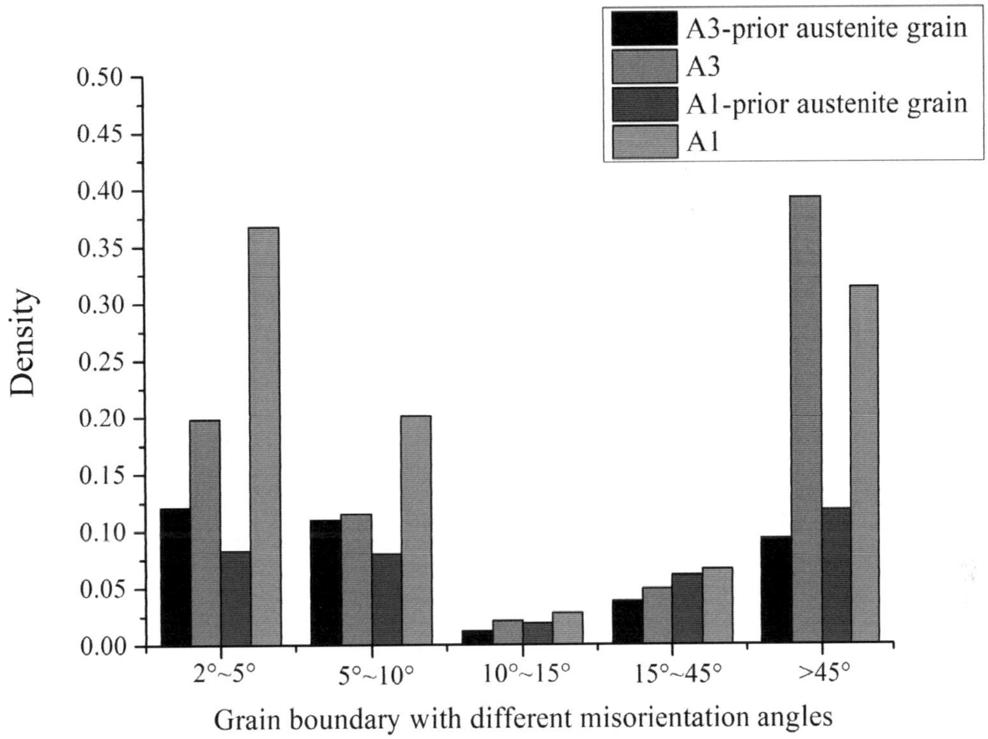

Figure 6 Density of grain boundaries with different misorientation angles in A1 and A3.

In this study, the 80mm ultra-heavy steel shows good quenching degree and major lath microstructure from the surface to the central layer, so the mechanical properties do not flunctuate a lot, as shown in Table 2, while the gradient microstructure also contributes to the change in strength and toughness. More >45° grain boundaries improve the toughness, so the higher >45° grain boundary density in the center of ultra-heavy steel leads to the higher toughness. With respect to the strength, both the effects of dislocation density and grain size are considered. The dislocation density could be calculated based on low-angle grain boundary density, and the grain size is related to >15° grain boundary density [24]. Higher low-angle grain boundary density in A1 increase the strength to higher extent, while the lower equivalent grain width contributes to the stronger strength improvement in A2 and A3. The coordination of two types of strength mechanism shows balanced effect on the through-thickness strength. In other words, the difference in strength is decreased along the thickness direction.

Table 2 Mechanical properties at different thickness.

	-84°C Impact Energy (MPa)	Yield Strength (MPa)	Tensile Sstrength (MPa)
A1(near surface)	166	835	906
A2 (1/4 layer)	176	825	899
A3(central layer)	190	827	899

4. Conclusions

1) The 80mm ultra-heavy steel shows major lath-structure, while lath morphology and grain boundary character distribution difference are displayed along the thickness. Compared with the surface layer,

more Σ3 grain boundaries are achieved in the 1/4 and central layers and lead to higher high-angle grain boundary density and lower low-angle grain boundary density, which corresponds to the thinner and longer grain morphology.

2) The gradient microstructure shows the effects of cooling rate and prior austenite grain size on variant selection during phase transformation. Moderate cooling rate in inner layers is beneficial for V1/V2 variant pairs and higher Σ3 grain boundary between the pairs. In contrast, the higher cooling rate in surface layer enhances the forming ability of V1/V4 variant pairs and corresponding 5°-10° boundaries. In addition, the higher cooling rate and smaller original austenite grain size simultaneously lead to more nucleation sites during phase transformation and thus contributes to stronger strain incompatibility, which is correspondent to more low-angle boundaries.

3) More low-angle boundaries in the surface layer and high-angle boundaries in inner layers contribute to stronger dislocation strengthening and grain boundary strengthening respectively, realizing the through-thickness property coordination to some extent. With respect to the toughness, the increasing >45° grain boundary density from the surface to the center of heavy steel leads to the gradual increase of toughness.

Acknowledgement: This work was supported by the CISRI Innovation Fund (20161761ZD).

References

[1] Fang Y P, Shang C J, Guo H, Misra R D K, Xie Z J and Han G 2014 Structure-property relationship in a 960MPa grade ultrahigh strength low carbon niobium-vanadium microalloyed steel: The significance of high frequency induction tempering *Mater. Sci. Eng.* A **618** 112-7
[2] Fang Y P, Xie Z J and Shang C J 2014 Effect of induction tempering on carbide precipitation behavior and toughness of a 1000 MPa grade high strength low alloy steel *Acta Metall. Sin.* **50** 1413-20
[3] Wu B B 2019 Study on crystallographic characteristics of high strength low alloy steel and its composition-process-performance relationship (University of Science and Technology Beijing: Beijing)
[4] Yuan S F 2020 Study on the microstructure controlling and mechanical properties of high-performance offshore engineering steel (University of Science and Technology Beijing: Beijing)
[5] Wu B B, Huang S, Wang Z Q, Zhao J X, Wang C S, Shang C J and Misra R D K 2019 Crystallography analysis of toughness in high strength ultra-heavy plate steel *Mater. Lett.* **250** 55-9
[6] Tao Z, Hao Y and Wang S 2016 Effect of microstructural types on toughness and microstructural optimization of ultra-heavy steel plate: EBSD analysis and microscopic fracture mechanism *Mater. Sci. Eng.* A **658** 150-8
[7] Morsdorf L, Jeannin O, Barbier D, Mitsuhara M, Raabe D and Tasan C C 2016 Multiple mechanisms of lath martensite plasticity *Acta Mater.* **121** 202-14
[8] Morito S, Huang X, Furuhara T, Maki T and Hansen N 2006 The morphology and crystallography of lath martensite in alloy steels *Acta Mater.* **54** 5323-31
[9] Fan L, Zhou D H, Wang T L, Li S R and Wang Q F 2014 Tensile properties of an acicular ferrite and martensite/austenite constituent steel with varying cooling rates *Mater. Sci. Eng.* A **590** 224-31

[10] Luo H W, Wang X H, Liu Z B and Yang Z Y 2020 Influence of refined hierarchical martensitic microstructures on yield strength and impact toughness of ultra-high strength stainless steel *J. Mater. Sci. Technol.* **51** 130-6

[11] Swarr T E and Krauss G 1976 The effect of structure on the deformation of as-quenched and tempered martensite in a Fe-0.2C alloy *Metall. Trans.* A **7** 41-8

[12] Morito S, Yoshida H, Maki T and Huang X 2006 Effect of block size on the strength of lath martensite in low carbon steels *Mater. Sci. Eng.* A **438-440** 237-40

[13] Xu Z Y 2009 Effect of lath martensite morphology on the mechanical properties of steel *HT* **24** 1-5

[14] Ma Y L, Liu Y, Zhang L P, Jiang X and Liu C M 2020 Effect of cooling rate on microstructure and mechanical properties of CB2 tempered martensitic steel *China Foundry* **17** 102-10

[15] Zhang C, Wang Q, Ren J, Li R, Wang M, Zhang F and Sun K 2012 Effect of martensitic morphology on mechanical properties of an as-quenched and tempered 25CrMo48V steel *Mater. Sci. Eng.* A **534** 339-46

[16] Gao G, An B, Zhang H, Guo H, Gui X and Bai B 2017 Concurrent enhancement of ductility and toughness in an ultrahigh strength lean alloy steel treated by bainitebased quenching-partitioning-tempering process *Mater. Sci. Eng. A* **702** 104–12

[17] Xia D X, Wang X L, Li X C, You Y and Shang C J 2013 Properties and microstructure of third generation X90 pipeline steel X90 *Acta Metall. Sin.* **49** 271-6

[18] Du C, Hoefnagles J P M, Vaes R and Geers M G D 2016 Block and sub-block boundary strengthening in lath martensite *Scripta Mater.* **116** 117-21

[19] Morito S, Tanaka H, Konishi R, Furuhara T and Maki T 2003 The morphology and crystallography of lath martensite in Fe-C alloys *Acta Mater.* **51** 1789-99

[20] Lambert P A, Gourgues A F and Pineau A 2004 Austenite to bainite phase transformation in the heat-affected zone of a high strength low alloy steel *Acta Mater.* **52** 2337-48

[21] Morito S, Pham A H, Hayashi T and Ohba T 2015 Block boundary analyses to identify martensite and bainite *Mater. Today: Proc.* **2S** S913-6

[22] Mirzaei A, Ghaderi R, Hodgson P D, Ma X, Rohrer G S and Beladi H 2022 The influence of parent austenite characteristics on the intervariant boundary network in a lath martensitic steel *J. Mater. Sci.* **57** 8904–23

[23] Takaki S, Fukunaga K, Syarif J and Tsuchiyama T 2004 Effect of grain refinement on thermal stability of metastable austenite steel *Mater. Trans* **45** 2245-51

[24] Zhang Z Y, Chai F, Luo X B, Chen G, Yang C F and Su H 2019 The Strengthening Mechanism of Cu Bearing High Strength Steel As-Quenched and Tempered and Cu Precipitation Behavior in Steel *Acta Metall. Sin.* **55** 783-91

A polyurethane solid-solid composite phase change material based on modified graphene oxide for efficient thermal management

X Yang[1,2], Y N Liao[1,2], J Li[1,2 *], Y Mao[1,2], S S Chen[1,2], K Sun[1,2], M Li[1,2] and Y J Zheng[1,2]

[1] Key Laboratory of Low-grade Energy Utilization Technologies and Systems, Ministry of Education, Chongqing University, Chongqing 400044, China

[2] School of Energy and Power Engineering, Chongqing University, Chongqing, 400044, China

* Corresponding author's e-mail: lj202740@cqu.edu.cn

Abstract. Solid-solid phase change materials (SSPCMs) are potential materials for energy storage and thermal management, and improving mechanical property and heat transfer performance is the key to practical application. After modified graphene oxide (TGO) was prepared by toluene diisocyanate, a kind of super-tough, form-stable and leak-proof polyurethane composite phase change material (PU-TGO) was constructed by the two-step polymerization of isophorone diisocyanate and 1, 4-butanediol using polyethylene glycol and TGO. TGO has more uniform and stable dispersion than traditional graphene oxide, and its chemical crosslinking with polyurethane phase change materials makes the composite have better shape stability, mechanical strength, thermal conductivity and flexibility. PU-TGO has thermal conductivity and elongation at break up to 0.3072 W/(m•K) and 540%, respectively, and maintains excellent thermal stability and 72.09 J/g enthalpy after 500 thermal cycles. PU-TGO prepared by chemical crosslinks in this paper further demonstrates the great potential of phase change materials in the field of thermal management of new electronic devices.

Key words: Phase change materials, modified graphene oxide, polyurethane, mechanical property, thermal conductivity

1. Introduction

As modern equipment shows the tend of small size and high energy density, the huge heat flux rapidly increases the temperature of devices such as batteries thus affecting their service performance, even causing fatal damage [1-3]. Therefore, good thermal management methods have become the focus of research [4-6]. Conventional heat dissipation methods include active cooling (air cooling [7, 8], liquid cooling [9, 10]) and passive cooling [11]. For small space heat dissipation problems, active cooling requires more energy, space and materials to support corresponding heat dissipation facilities. In comparison materials used for passive cooling, such as phase change materials (PCMs), can absorb and release a large amount of latent heat at almost stable temperature points [12], and their excellent heat storage capacity and controllable material shape better match the heat dissipation space of the equipment, achieving low energy consumption and low cost heat absorption and storage, thereby ensuring that equipment can operate within a safe and reasonable temperature range [13-16].

PCMs can be divided into organic and inorganic types. Inorganic phase change materials have not yet been widely used due to the problems of phase separation, supercooling, and corrosivity [17]; while organic phase change materials such as polyethylene glycol (PEG) [18, 19], paraffin [20, 21], and fatty acids [22, 23] can avoid these problems, but they also have shortcomings such as low thermal conductivity, poor mechanical properties, and leakage [24, 25].

Many studies have used high thermal conductivity fillers to improve the thermal conductivity of phase change materials, such as expanded graphite [26], graphene oxide (GO) [27], carbon nanotubes [28], SiO_2 [29], Cu nanoparticles [30], boron nitride [31]. The surface of GO contains a large number of functional groups that are easy to react or bond with other organic matter[32], and has small volume and high thermal conductivity, so it has become one of the most popular nanofillers. Zhang et al. [33] established graphene aerogel (GA) by introducing PEG into GO, and prepared a shape stable GA/PEG composite phase change material by impregnation. Benefiting from the capillary force, support effect, and high thermal conductivity channels of GA, the thermal conductivity of 4.6wt% PEG-10K, PEG-1K, and PEG-30K increased by about 28.86%, 27.74%, and 33.26%, respectively, compared to pure PEG. Safaei et al.[34] dispersed GO with phi = 0.2, 0.4 and 0.6 wt.% in paraffin wax to prepare composite PCMs for desalination. The results showed that the addition of GO can reduce the melting temperature of paraffin wax and increase the productivity of solar distillers by 25%.

The modification of GO can make it have the ability to form chemical bonds with PCM systems[35, 36]. Pielichowska et al.[37] prepared polyurethane-based PCMs containing polyethylene glycol fragments with an average molar mass of 8000 g/mol, and found that the phase change heat and crystallinity of the 0.3% graphite nanosheet modified system were the highest in polyurethane without chain extender. Graphite nanosheets change the phase transition behavior and crystallinity of polyurethane matrix, and the thermal conductivity of polyurethane materials increases with the increase of graphite nanosheet content.

Furthermore, the mechanical strength and leakage resistance of composite materials is often improved by using microcapsule shell-core structure [38-41] and skeleton adsorption [42-45]. However, these strategies which can maintain high shape stability and thermal conductivity, still couldn't overcome the brittleness and the lack of flexibility. Chemical synthesis methods improve the mechanical properties of PCMs while preventing leakage, not only relying on brittle skeletons and hard shells. Moreover, through the design of molecular structures, the chemical crosslinking provides a certain degree of flexibility and overall strength for the final PCM such as polyurethane based PCMs to adapt to a variety of working environments and scenarios [46]. Due to the stable properties, high phase change enthalpy, easy modification and chemical performance, polyurethane-based PCMs with solid-solid phase change process have also received more and more attention. Liang et al.[47] synthesized a series of thermoplastic polyurethane-based solid-solid PCMs by using the three-step condensation reaction of polyethylene glycol with TABE, 4,4'-diphenylmethane diisocyanate and polyethylene glycol monomethyl ether. The maximum enthalpy of melt and crystallization of this material is 153.51 J/g and 143.91 J/g, respectively. The measured initial decomposition temperature is 320.16 °C, and the optimal phase transition temperature is between 22.41 ~ 48.41 °C.

In this work, a new synthesis method was proposed: GO with improved dispersion in solution was used as a nanofiller with high thermal conductivity, and introduced it into a chemical cross-linking network composed of IPDI, PEG, and BDO through modification, discarding the disadvantage of traditional physical addition of GO with low interfacial strength. The designed synthesis strategy can simultaneously improve the thermal conductivity and mechanical properties of phase change materials. The polyurethane based composite phase change material has excellent thermal stability and a certain degree of flexibility, without macroscopic leakage during 500 thermal cycles, and its ultimate tensile ratio can reach over 500%. The successful synthesis and reproducibility of this material will further promote the practical application of phase change materials and expand their potential in chemical synthesis.

2. Experimental

2.1 Materials

Polyethylene glycol (PEG, Mn = 4000 Da), Dibutyltin dillaurate (DBDTL), Molecularsieve (5Å), 2mm-3mm, Toluene diisocyanate (TDI) were purchased from Shanghai Aladdin Biochemical Technology Company (China). Isophorone diisocyanate (IPDI) and N, N-dimethylformamide (DMF) were purchased from Shanghai Macklin Biochemical Company (China). 1, 4-Butanediol (BDO) were purchased from Chengdu Kelong Chemical Reagent Company (China). Graphene oxide (GO) were purchased from Shanghai Yuanye Bio-Technology Company (China).

2.2 Modification of GO

Take 0.5 g GO in a three-necked flask, add 25 ml anhydrous DMF, cover with a rubber stopper, disperse ultrasonic for 0.5 h, then put it in an oil bath, stir to a constant temperature 75°C under the protection of nitrogen. Extract 4.067 ml of TDI and mix it in a glass bottle with 16.65 ml of anhydrous DMF, add 5 drops of catalyst, and react for 5 h. Centrifuge several times until the supernatant is clear. The obtained product was soaked in anhydrous DMF for 12 h to wash away the unreacted TDI on the surface. Titration of the filtrate confirmed no TDI residue. Place the product in a vacuum drying oven at 45°C to dry for 12 h, grind into powder for later use.

2.3 Synthesis of polyurethane composite PCM

Add 21 g PEG4000 in a three-necked flask and vacuum dry at 100°C for 3 h, cool down to 80°C and introduce nitrogen. Weigh 0.0144 g (0.06%) of DBDTL and 2.5 g IPDI, 0.1 g GO/TGO and 3 ml DMF miscible in small glass bottles and co-inject into the flask. Mechanical stirring at 80°C nitrogen environment for 1 h, cool down to 70°C, weigh 0.4506 g BDO and 10 ml DMF in a small glass bottle, mix well, then co-inject into the flask, mechanical stirring at 70°C nitrogen environment for 2 h. Pour into a PTFE Petri dish and air dry to form a film at room temperature.

2.4 Characterization

The fourier transform infrared (FTIR) spectras were performed by Nicolet iS50 infrared spectrometer (Thermo fisher Scientific, USA). The thermal stability was measured by thermogravimetric analyzer (TGA2, METTLER TOLEDO, Switzerland). Differential scanning calorimetry (DSC) measurements were obtained on DCS404 F3 (Netzsch, Germany). X-rays diffraction (XRD) analysis was presented by powder diffractometer (Panalytical X' Pert, The Netherlands). The stress-strain curve was gathered by universal tensile tester (ETM102B, China), each sample is made into a barbell shape with a tensile rate of 10mm/min. Thermal conductivity were measured by TPS 2500S thermal conductivity meter (Hot Disk, Sweden). Thermal stability was investigated by heating the samples in an oven at 80°C for 20 min then cool them to 20°C for 20 min, and so on for 500 cycles. The leakage tests were carried through heating the samples from 20°C to 40°C and 90°C for 1 h to observe the leakage behavior.

3. Results and discussion

3.1 TDI-modification of GO

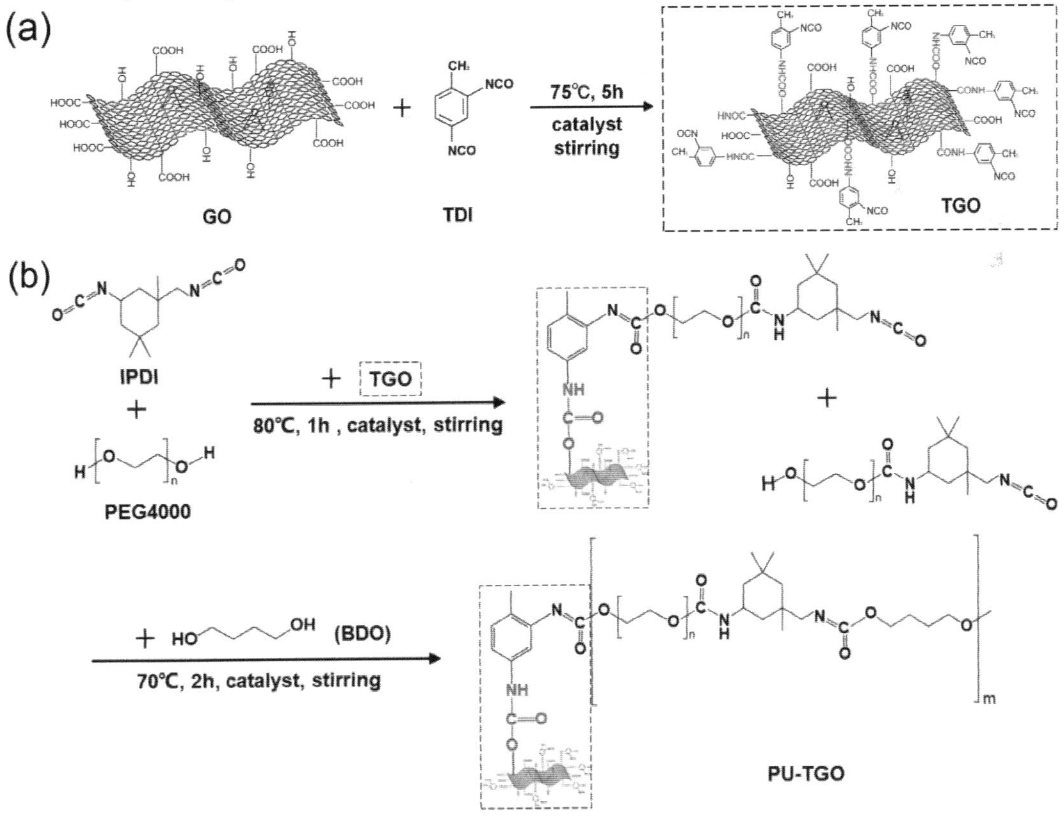

Figure 1. Synthesis of (a) TGO and (b) PU-TGO.

The dispersion of graphene oxide [27, 48-50] is an important factor affecting the properties of composite materials, and chemical modification can further improve its dispersion degree and stability. In this study, graphene oxide was modified using TDI to improve its dispersion in solvents (Figure 1a). TDI modification enables GO to form chemical crosslinks with polyurethane phase change materials. The two-step method is used to graft IPDI, PEG4000, TGO and BDO to form PU-TGO long-chain polymers (Figure 1b), mainly involving the reaction of hydroxyl groups and isocyanate groups.

In order to compare the difference between dispersions of TGO and GO, dissolved equal amount of TGO and GO in the organic solvent DMF and mixed them homogenously by ultrasound treatment (Figure 2a). After 24 hours, GO occurred significant stratification in solution, while the TGO in the other centrifuge tube remained stable and dispersed despite a small amount of precipitation. After 48 h of sonication, the GO was almost completely separated from the DMF and precipitated at the bottom of the centrifuge tube, while the vast majority of TGO was scattered in DMF as it was at 24 hours. The result confirmed that TGO has better dispersion than GO, which is favorable for the preparation of composite PCMs with better uniformity in subsequent experiments, which is due to the oxygen-containing groups such as epoxy, hydroxyl, carbonyl and carboxyl groups contained in graphene oxide, making it hydrophilic[51-53]. Whereas TDI reacted with the surface oxygen-containing groups of GO during the modification process, the benzene rings introduced after reaction and the formed isocyanate groups are hydrophobic groups, which reduced the hydrophilic properties of GO[54]. In addition, the long chain on the surface of TGO prevents direct contact between graphene sheets to a certain extent. Therefore, TGO can be better dispersed in organic solvents.

The chemical structure of GO and TGO were verified by FTIR spectra (Figure 2b). The expansion vibration absorption peaks of hydroxyl (-OH) and carboxyl groups (-COOH) on the GO surface were located at 3341cm-1 and 1721cm-1, and the carbonyl (C=C), hydroxyl (-OH) and ether groups (C-O-C) telescopic vibration absorption peaks on the GO skeleton were located at 1611 cm-1, 1220cm-1 and 1045 cm-1, respectively. Compared to GO, the FTIR spectra of TGO shows new absorption peaks of N-H, C=O and C-N corresponding to the telescopic vibration absorption peaks of urethane groups (RNHCOOR') or amide groups (RCONH2) on TGO backbones. Meanwhile, a new telescopic vibration absorption peak appeared at TGO at 2287 cm-1, corresponding to the TDI para-unreacted isocyanate group, and the expansion vibration absorption peaks of TDI's own benzene ring and methyl group were also at 1601 cm-1 and 1380 cm-1 in TGO infrared spectroscopy, respectively. In summary, GO was successfully modified by TDI, and one isocyanate group of TDI reacts with oxygen-containing functional groups on the surface of GO, and the other remains at the end of the chain [47].

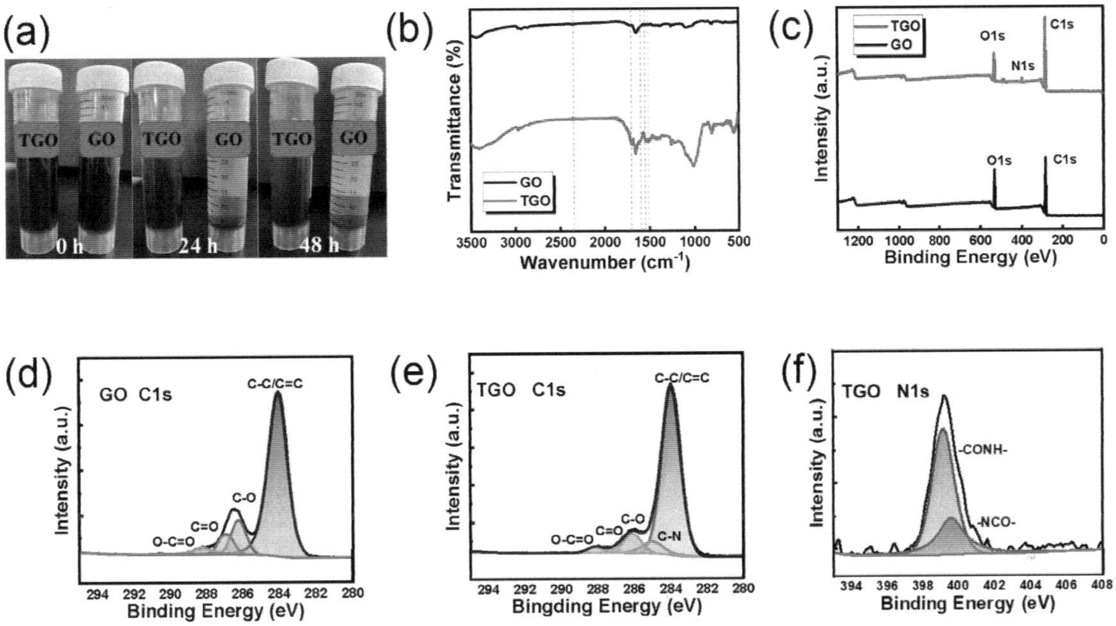

Figure 2. (a) Dispersion experiment over time of TGO and GO in DMF. (b) FITR spectra of GO and TGO. (c) XPS full spectrum of TGO and GO. (d) C1s spectrogram of GO. (e) C1s spectrogram of TGO. (f) N1s spectrogram of TGO.

To further explore the changes of GO surface elements before and after modification, C, N and O contents of GO and TGO were characterized by XPS (Figure 2c and Table 1). After TDI modification, the O1s peak intensity of GO decreased, and the oxygen content decreased from 19.34% to 12.82%. Meanwhile, a new N1s absorption peak appeared at 400 eV, and the nitrogen content increased from 1.19% to 5.04%. Compare the C1s spectrograms of GO (Figure 2d) and TGO (Figure 2e), the C-O bond of TGO is weakened and a new bond C-N appeared in the 286 eV bound state, because the oxygen-containing functional groups on the GO surface reacted with isocyanates to form urethane groups (RNHCOOR') or amide groups (RCONH2). The N1s spectra of TGO (Figure 2f) include 399.5 eV and 400.5 eV binding states, corresponding to -CONH-, -NCO-, respectively, further supporting that one isocyanate group of TDI reacted with GO and the other remains on the surface of GO through grafting.

Table 1. C、N and O contents of GO and TGO.

Samples	C (at.%)	O (at.%)	N (at.%)
GO	79.47	19.34	1.19
TGO	82.14	12.82	5.04

3.2 Phase change properties of polyurethane composite PCMs

The polyurethane composite PCMs prepared by GO and TGO are black films with uniform thickness adjustment, which can be bent and folded arbitrarily, and the surface didn't leave any traces after bending, with excellent flexibility (Figure 3a).

The thermophysical properties of PU-GO and PU-TGO are shown in Table 2. Unreacted isocyanate group of TDI during modification continued to react with the hydroxyl group of PEG4000, reducing the effective crystallization region of PEG4000, resulting in a decrease in the crystal enthalpy of PU-TGO compared with PU-GO. At the same time, due to the reaction of isocyanate groups on the surface of PU-TGO with the formation of defective crystal nuclei, the melting enthalpy of PU-TGO is somewhat lower than that of PU-GO, and the melting temperature is also slightly reduced.

The XRD spectra of PU-GO, PU-TGO and PEG4000 (Figure 3b) showed that the introduction of GO and TGO did not change the position of PEG4000-derived peaks, and the main crystalline segment is still polyethylene glycol, but with reduced peak intensity, which proved that the isocyanate group of TDI reacted with the hydroxyl group of PEG4000 and reduced its crystallization strength.

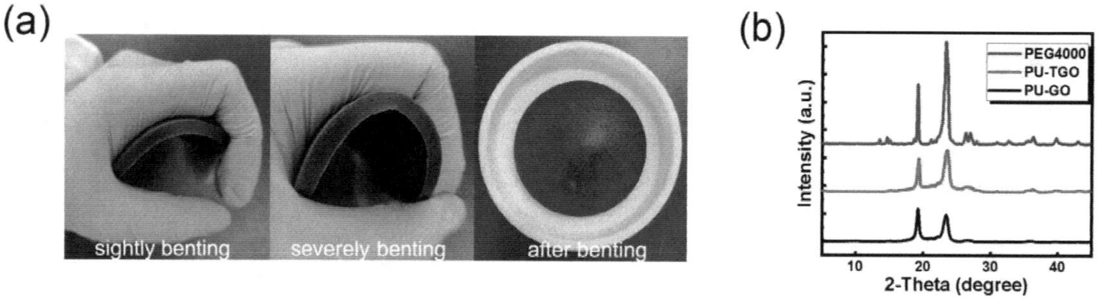

Figure 3. (a) Optical images of polyurethane composite PCM. (b) XRD spectra of PU-GO, PEG4000 and PU-TGO.

The optical leakage tests at 20°C, 40°C and 90°C for 1 hour were adopted to evaluate the shape stability of PU-GO and PU-TGO (Figure 4a). Pristine PEG4000 had obvious liquid leakage penetration filter paper at 40°C and completely leaked at 90°C. In contrast, PU-GO and PU-TGO had never leaked, proving excellent shape stability. The surface image of PU-GO at 20°C proved that the unmodified GO was physically compounded in the polyurethane system, and white is the color of the polyurethane PCM crystal, while the better dispersibility and chemical crosslinking of TGO makes PU-TGO always black.

3.3 Thermal stability and thermal cycling performance of polyurethane composite PCMs

Thermogravimetric Analysis (TGA, Figure 4b) and Derivative Thermogravimetry (DTG, Figure 4c) results confirmed the high thermal stability of PU-GO and PU-TGO. Both of them begin to thermally decompose at 283°C, which is much higher than the ideal working environment below 200°C. The initial thermal decomposition temperature of pristine PEG is 400°C, the chain expansion reaction generates a large number of urethane groups, and most of the oxygen-containing functional groups on the surface of GO decompose into oxygen-containing free radicals below 100°C, which ultimately reduces the initial thermal decomposition temperature and the stability of the whole system.

The thermal reliability of PU-GO and PU-TGO was characterized by differential scanning calorimeter (DSC) analysis (Figure 4d). The phase change temperatures and other specific parameters are shown in Table 2. After 500 thermal cycles of PU-GO and PU-TGO, the melting enthalpy and crystal enthalpy decreased slightly, but still maintained good heat storage performance within the acceptable range. The polyurethane composite PCMs still maintain good heat storage capacity (more than 70 J/g) and stability after 500 thermal cycles, and no macroscopic leakage occurred in 500 thermal cycles, which proved the excellent thermal cycle stability of PU-GO and PU-TGO. The phase change enthalpy of PU-TGO is lower than that of PU-GO since chemical crosslinking reduces the effective crystallization segment of PEG, but the difference between the two is not large. The change of phase change characteristic of PU-TGO after cycling is smaller, which proves that PU-TGO has better thermal stability than PU-GO.

Table 2. Thermal physical properties of PU-GO and PU-TGO before and after 500 thermal cycles[a].

Samples	H_m (J/g)	T_m (°C)	H_c (J/g)	T_c (°C)
PU-GO	89.84	40.65	81.81	25.11
PU-TGO	85.13	38.32	77.57	25.05
PU-GO-500	82.85	47.83	75.48	30.38
PU-TGO-500	73.14	40.14	72.09	23.19

[a] Hm, Hc, Tm, and Tc are the melting enthalpy, crystal enthalpy, melting temperature, and crystal temperature of polyurethane composite PCMs, respectively.

3.4 Mechanical properties and thermal conductivity of polyurethane composite PCMs

Mechanical properties of PU-GO and PU-TGO were characterized by tensile tests (Figure 4e and Table 3). The mechanical properties of pristine PEG4000 are not shown due to its poor range, moreover, the elongation at break, tensile strength and toughness of PU-TGO are higher than those of PU-GO. This is because TGO has high dispersion in polyurethane-based phase change materials. Moreover, the chemical crosslinking between the functional groups modified on the surface and the oxygen-containing functional groups of the PCM and chain extender also enhances the interface strength between TGO nanoparticles and the material, then the uniformly dispersed TGO can undertake part of the external force and improve the deformation resistance of the entire composite phase change material.

Table 3. Mechanical parameters of PU-GO and PU-TGO.

Samples	Strain(%)	Stress (MPa)	Toughness (MJ/m^3)
PU-GO	499	16.1	83.7
PU-TGO	540	19.6	100.3

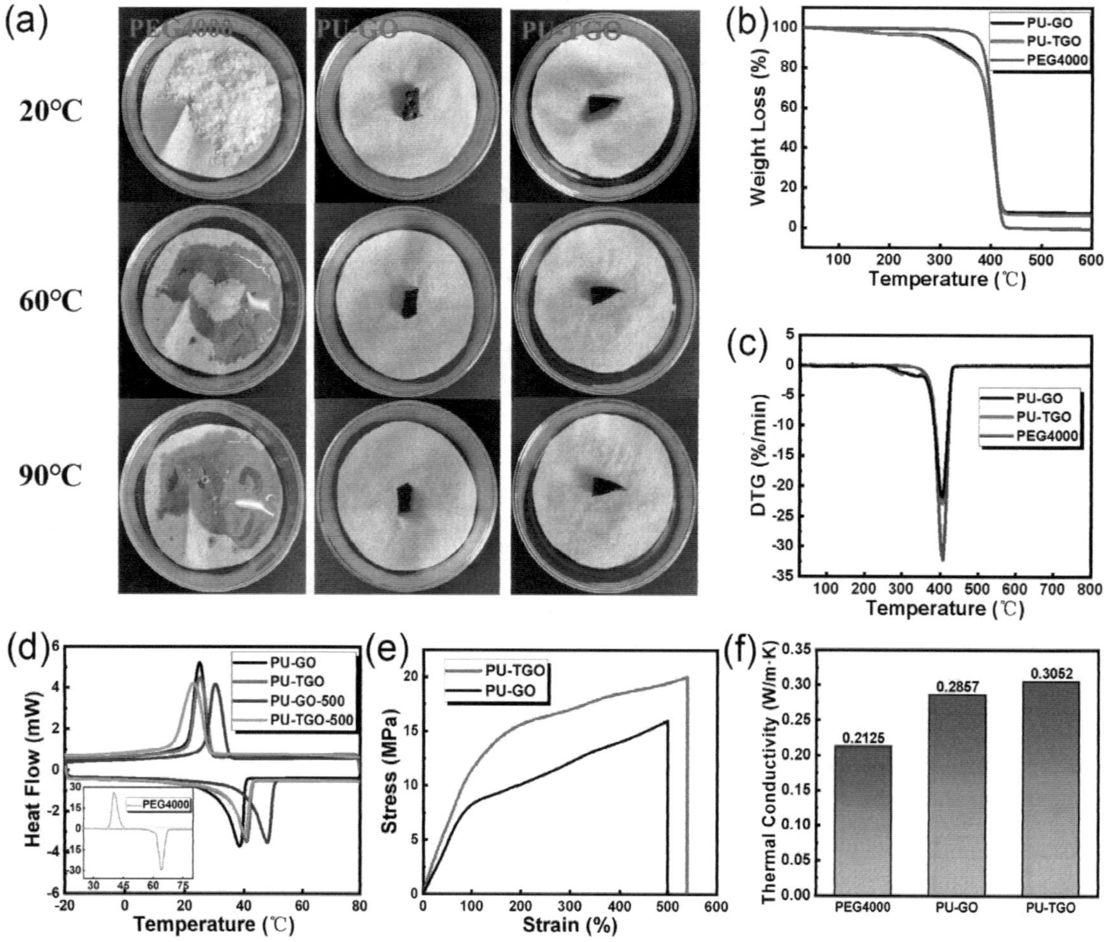

Figure 4. (a) Optical images of PEG4000, PU-GO and PU-TGO placed at 20°C, 60°C, 90°C for 1 hour. (b) TGA curves and (c) DTG curves of PU-GO, PU-TGO and PEG4000. (d) DSC curves of PU-GO and PU-TGO before and after 500 thermal cycles (Illustrated DSC curve of PEG4000). (e) Stress-strain curves of PU-GO and PU-TGO. (f) Thermal conductivity of PEG4000, PU-GO and PU-TGO.

The results of the thermal conductivity test show that the introduction of GO and TGO increases the thermal conductivity of the phase change material by 34% and 45%, respectively (Figure 4f). This is due to the high thermal conductivity of GO particles, and TGO can react with the substrate to form chemical links compared to GO, so that the material as a whole has higher interface strength and a more continuous thermal conductivity path, then the thermal conductivity of PU-TGO is higher.

4. Conclusions

In summary, a polyurethane solid-solid composite PCM based on modified GO was synthesized by combining GO, IPDI, BDO and PEG4000 through a two-step chemical polymerization reaction. In PU-TGO system, GO is enhanced in dispersion in the system after TDI modification, and chemically reacts with other materials to form bonds. The addition of GO improves the mechanical properties, leakage resistance and thermal stability of composite phase change materials. Due to the good mechanical and thermal properties of polyurethane PCM, PU-GO also has a good performance. However, the chemical crosslinking of TGO further improves the interfacial strength, stability and heat transfer capacity of the composites, and at the same time affects the crystallization, resulting in a slight decrease in the phase change enthalpy, but the defects cannot obscure the virtues.

PU-GO and PU-TGO can withstand the stress of 16.1 MPa and 19.6 MPa respectively, and have good shape stability, and maintain good heat storage capacity after 90°C and 500 thermal cycles. Compared with the unmodified PU-GO, the internal chemical cross-linking of PU-TGO forms a stronger interfacial bonding strength and a more continuous heat transfer path. Therefore, PU-TGO has better elongation and thermal conductivity up to 540% and 0.3072w/m • K, respectively. The flexible super-rigid polyurethane composite PCM developed in this study has great potential in the heat dissipation of electronic devices, and can even provide certain mechanical support for devices.

Acknowledgments

The work was financially supported by National Key R&D Program of China (2022YFB3803300), Chongqing Young Talents Support Program (CQYC2021059206), National Natural Science Foundation of China (51606017), Special Fund for Basic Scientific Research Funds of Central Universities (2020CDJQY-A055), Chongqing Outstanding Youth Fund (cstc2021jcyj-jqX0015).

References

[1]Li Tan, Li Zhang, Qingna Sun, Ming Shen, Qunting Qu and Honghe Zheng 2013 Electrochimica Acta 111 802-808

[2]Todd M. Bandhauer, Srinivas Garimella and Thomas F. Fuller 2011 Journal of the Electrochemical Society 158 R1-R25

[3]S. S. Anandan and V. Ramalingam 2008 Thermal Science 12 5-26

[4]Jaewan Kim, Jinwoo Oh and Hoseong Lee 2019 Applied thermal engineering 149 192-212

[5]Hussam Jouhara, Navid Khordehgah, Nicolas Serey, Sulaiman Almahmoud, Stephen R. Lester, Daniel Machen and Luiz Wrobel 2019 Energy 170 849-861

[6]Antti Lajunen, Yinye Yang and Ali Emadi 2020 Ieee Transactions on Vehicular Technology 69 6025-6040

[7]Yun Gao, Wujun Ji and Xiaoqiang Chen 2022 Processes 10 1104

[8]Seham Shahid and Martin Agelin-Chaab 2018 International Journal of Energy Research 42 2246-2262

[9]Wei Tang, Hua Ding, Xiaoming Xu, Jiaqi Fu, Lin Liu, Wang Wei, Ye Xiao and He Huang 2020 Journal of Renewable and Sustainable Energy 12 045701

[10]Jin-Kuk Kim and Robin Smith 2001 Chemical Engineering Science 56 3641-3658

[11]Jinyong Kim, Chuanbo Yang, Joshua Lamb, Andrew Kurzawski, John Hewson, Loraine Torres-Castro, Anudeep Mallarapu and Shriram Santhanagopalan 2022 Journal of the Electrochemical Society 169 110543

[12]Xuan Tong, Nianqi Li, Min Zeng and Qiuwang Wang 2019 Renewable & Sustainable Energy Reviews 108 398-422

[13]Dong-Lin Han, Lu-Yao Yang, Shuai-Peng Wang, Xian-Qing Zeng, Lei Tang, Kai Liu, Kun Liang, Hong-Qian Shen-Tu, Lin-Qing Guo, Lu Bai andYu-Chuan Huang 2022 Journal of Materials Science-Materials in Electronics 33 1008-1020

[14]Yanan Wang, Zhengkun Wang, Haitao Min, Hua Li and Qingfeng Li 2021 Journal of Energy Storage 35 102279
[15]Malek Nofal, Said Al-Hallaj and Yayue Pan 2020 Applied Thermal Engineering 171 115126
[16]Guiwen Jiang, Juhua Huang, Yanshu Fu, Ming Cao and Mingchun Liu 2016 Applied Thermal Engineering 108 1119-1125
[17]Kunjie Yuan, Jinming Shi, Waseem Aftab, Mulin Qin, Ali Usman, Fang Zhou, Yang Lv, Shan Gao and Ruqiang Zou 2020 Advanced Functional Materials 30 1904228
[18]Shaowei Li, Jing Li, Yanning Liao, Shanshan Chen, Yujie Zheng, Meng Li and Kuan Sun 2022 Journal of Energy Storage 54 105337
[19]H. Li and G. Y. Fang 2010 Chemical Engineering & Technology 33 1650-1654
[20]Jintao Lu, Tao Zhang, Zhiguang Xu and Yan Zhao 2021 Chemical Engineering Journal 425 131432
[21]Yuang Zhang, Malik Muhammad Umair, Shufen Zhang and Bingtao Tang 2019 Journal of Materials Chemistry A 7 22218-22228
[22]Quanying Yan, Chao Liu and Jing Zhang 2019 Materials Research Express 6 065507
[23]A. Sari, A. Karli, C. Alkan and A. Karaipekli 2013 Energy Sources Part a-Recovery Utilization and Environmental Effects 35 1813-1819
[24]Malik Muhammad Umair, Yuang Zhang, Kashif Iqbal, Shufen Zhang and Bingtao Tang 2019 Applied Energy 235 846-873
[25]Nan Zhang, Yanping Yuan, Xiaoling Cao, Yanxia Du, Zhaoli Zhang and Yewei Gui 2018 Advanced Engineering Materials 20 1700753
[26]Jian-guo Zhao, Quan-gui Guo, Xiao-qing Gao, Xin-hai Wei, Jing-li Shi, Li-zhong Yao and Lang Liu 2009 New Carbon Materials 24 114-118
[27]Zhifang Liu, Zhonghua Chen and Fei Yu 2019 Solar Energy Materials and Solar Cells 192 72-80
[28]Qianyun Cao, Fangfang He, Yinjun Li, Zhiyu He, Jinghui Fan, Renchao Wang, Wenxia Hu, Kai Zhang and Wenbin Yang 2020 Fullerenes Nanotubes and Carbon Nanostructures 28 656-662
[29]Jian-li Shang, Hao Zhang, Lei Xiong and Xiang-long Ma 2016 Spectroscopy and Spectral Analysis 36 2430-2436
[30]S. Y. Wu, H. Wang, S. Xiao and D. S. Zhu 2012 Journal of Thermal Analysis and Calorimetry 110 1127-1131
[31]Richa Agrawal, Joshua Hanna, I. Emre Gunduz and Claudia C. Luhrs 2019 Molecules 24 1883
[32]Sasha Stankovich, Dmitriy A. Dikin, Geoffrey H. B. Dommett, Kevin M. Kohlhaas, Eric J. Zimney, Eric A. Stach, Richard D. Piner, SonBinh T. Nguyen and Rodney S. Ruoff 2006 Nature 442 282-286
[33]Ping Zhang, Jiajun Li, Ruishi Xie, Jia Shen, Lixian Song and Lin Chen 2020 Diamond and Related Materials 103 107716
[34]Mohammad Reza Safaei, Hamid Reza Goshayeshi and Issa Chaer 2019 Energies 12 2002
[35]Guojia Huang, Zhigang Chen, Maodong Li, Bo Yang, Mingliang Xin, Shiping Li and Zongjie Yin 2016 Acta Chimica Sinica 74 789-799
[36]Yu Wang, SiSi Li, HaiYan Yang and Jie Luo 2020 Rsc Advances 10 15328-15345
[37]K. Pielichowska, J. Bieda and P. Szatkowski 2016 Renewable Energy 91 456-465
[38]Wenbin Yang, Li Zhang, Yongli Guo, Zhuoni Jiang, Fangfang He, Changqiong Xie, Jinghui Fan, Juying Wu and Kai Zhang 2018 Journal of Materials Science 53 2566-2575
[39]Li Zhang, Wenbin Yang, Zhuoni Jiang, Fangfang He, Kai Zhang, Jinghui Fan and Juying Wu 2017 Applied

Energy 197 354-363

[40]Jun-Feng Su, Xin-Yu Wang, Shan Han, Xiao-Long Zhang, Yan-Dong Guo, Ying-Yuan Wang, Yi-Qiu Tan, Ning-Xu Han and Wei Li 2017 Journal of Materials Chemistry A 5 23937-23951

[41]Xia Dai and Xiaodong Shen 2006 Rare Metals 25 393-399

[42]Dimberu G. Atinafu, Wenjun Dong, Changmin Hou, Radoelizo S. Andriamitantsoa, Jingjing Wang, Xiubing Huang, Hongyi Gao and Ge Wang 2019 Materials Today Energy 12 239-249

[43]Waseem Aftab, Asif Mahmood, Wenhan Guo, Muhammad Yousaf, Hassina Tabassum, Xinyu Huang, Zibin Liang, Anyuan Cao and Ruqiang Zou 2019 Energy Storage Materials 20 401-409

[44]Yanhong Wei, Juanjuan Li, Furong Sun, Jinrong Wu and Lijuan Zhao 2018 Green Chemistry 20 1858-1865

[45]Peng Min, Jie Liu, Xiaofeng Li, Fei An, Pengfei Liu, Yuxia Shen, Nikhil Koratkar and Zhong-Zhen Yu 2018 Advanced Functional Materials 28 1805365

[46]Qiqiu Huang, Xinxi Li, Guoqing Zhang, Yongchun Kan, Canbing Li, Jian Deng and Changhong Wang 2022 Applied Energy 309 118434

[47]Peng Liang, Peng Xi, BoWen Cheng, XiaoQing Wang, YaQing Zhang and Jiafeng Wu 2015 Science of Advanced Materials 7 2420-2426

[48]Vijayananth Kavimani, Balasubramaniam Stalin, Pudhupalayam Muthukutti Gopal, Manickam Ravichandran, Alagar Karthick and Murugesan Bharani 2021 Advances in Polymer Technology 2021 6627743

[49]Pei Chen, Fengwei Xie, Fengzai Tang and Tony McNally 2020 International Journal of Biological Macromolecules 158 420-429

[50]Mohammad Mehrali, Sara Tahan Latibari, Mehdi Mehrali, Teuku Meurah Indra Mahlia, Emad Sadeghinezhad and Hendrik Simon Cornelis Metselaar 2014 Applied Energy 135 339-349

[51]C. Nethravathi and Michael Rajamathi 2008 Carbon 46 1994-1998

[52]Suneel Kumar Srivastava and Juergen Pionteck 2015 Journal of Nanoscience and Nanotechnology 15 1984-2000

[53]Arunjunai Raja Shankar Santha Kumar, Francesco Piana, Matej Micusik, Juergen Pionteck, Susanta Banerjee and Brigitte Voit 2016 Materials Chemistry and Physics 182 237-245

[54]Peng Lin, Linghui Meng, Yudong Huang, Li Liu and Dapeng Fan 2015 Applied Surface Science 324 784-790

Molecular dynamics simulation of thermal properties of modified graphene / n-octadecane composite phase change material

J Li[1, 2*], Y Mao[1, 2], X Yang[1, 2], Y N Liao[1, 2], D Cai[1, 2], S S Chen [1, 2], K Sun [1, 2] and Y J Zheng [1, 2]

[1] Key Laboratory of Low-grade Energy Utilization Technologies and Systems, Ministry of Education, Chongqing University, Chongqing 400044, China

[2] School of Energy and Power Engineering, Chongqing University, Chongqing, 400044, China

* Corresponding author's e-mail: lj202740@cqu.edu.cn

Abstract. Phase change materials have been widely used in building energy-saving and off-peak energy storage systems, but most phase change materials limit their application because of their low thermal conductivity. Improving their heat transfer performance and revealing the heat transfer mechanism from a microscopic perspective are the keys to practical applications. The PCM system composed of n-octadecane and modified graphene was established by molecular dynamics simulation and the influence of macro thermal properties was analyzed from a microscopic perspective including end-to-end distance and torsional angle distribution, radial distribution function, and self-diffusion coefficient. The results show that the thermal conductivity of PCMS increases with increased modified graphene content. When the mass fraction is 26.88%, the thermal conductivity increases by 44.5%, which is consistent with the trend of experimental values. The addition of modified graphene resulted in a more concentrated distribution of n-octadecane molecules, indicating that the arrangement of n-octadecane molecules was affected, increasing the thermal conductivity of n-octadecane. The deviation of phase transition temperature from the experimental value is less than 1%.

1. Introduction

Solar energy, as a non-polluting and endless supply of energy, is considered to be the best choice to solve energy problems at present[1]. However, due to the intermittent and unstable energy supply of solar energy, thermal energy storage is considered one of the key technologies for the development of solar energy[2,3]. In addition, thermal energy is a low-grade energy source, which is often regarded as waste heat and difficult to use effectively[4]. Energy storage technology is a way to reduce the unevenness between energy supply and demand through energy storage, thereby improving efficiency and reducing the cost of renewable energy technologies[5]. Among various thermal energy storage methods, latent heat storage based on phase change materials (PCMs) is the most effective technology[6,7]. PCM has many excellent characteristics, such as extremely high heat storage density, strong chemical stability, and low cost[8,9]. It can be used in industrial waste heat recovery, solar energy systems, and construction[10–13]. PCM has almost no temperature change during phase change, and can be used for temperature control[14–16].

The low thermal conductivity is the most common disadvantage of most PCMs, which limits the thermal response rate of thermal energy storage systems[17]. Carbon-based materials, including graphene[18,19], expanded graphite[20], and multi-walled carbon nanotubes[21,22], are the most promising high thermal conductivity fillers[23]. Due to their good thermal conductivity, they can be used to enhance the heat transfer of phase change heat storage materials.

Graphene has extremely high intrinsic thermal conductivity, large specific surface area, and excellent chemical stability. Based on previous research, it was confirmed that adding nanographene sheets to n-octadecane can effectively enhance its thermal conductivity. When the mass fraction of nanographene sheets is 2wt.%, the thermal conductivity of phase change materials increases by 89.4%[24].

Although composite PCMs have good application prospects, cyclic stability is still a problem[25]. In composite PCM systems, graphene materials will agglomerate and settle due to the van der Waals force, resulting in a decrease in thermal conductivity. Although surfactants can be added to inhibit agglomeration[26], they may affect the physical properties of composite PCMs. Some studies have strengthened the dispersibility in EG by grafting functional groups on the edges of graphene[27]. Our previous experiments have shown that adding modified graphene (MG) to n-octadecane can greatly improve the energy storage efficiency of PCMs[28].

As an effective method for the microscopic analysis of materials, molecular dynamics methods have made great progress in the study of the thermophysical properties of materials. With the development of molecular dynamics methods, some meaningful explorations have appeared in the microscopic research of alkane PCMs. Sosso et al. used molecular dynamics software to simulate the rapid crystallization process of composite PCMs[29]. Yin et al. used molecular dynamics simulation to calculate the heat capacity and self-diffusion coefficient of the n-hexadecane system, and the freezing point of n-hexadecane was obtained accordingly[30]. Babaei et al. used molecular dynamics simulations to study the phase transition process of PCM systems with carbon nanotubes (CNTs) or graphene, and calculated the thermal conductivity of the system using balanced molecular dynamics simulations. The results shows that CNTs make the thermal conductivity of PCMs increased by 66% (solid) and 48% (liquid), graphene increased the thermal conductivity of PCM by 87% and 52%[31]. Rao et al. calculated the phase change behavior of alkane phase change materials, and the simulation results were close to the experiments, which indicating that the addition of water molecules into the alkane molecules would reduce the system fluidity and increase the heat storage capacity[32]. Zabihi et al. found that the functional groups of hydrogenation, methyl and phenyl on the surface of graphene were helpful to improve the thermal conductivity. As the coverage of H, CH_3 and C_6H_5 functional groups on the surface of graphene increased, the thermal resistance of the graphene-paraffin interface decreased, and the thermal conductivity of the nanocomposite increased[33]. Wang et al found that the hydrogenation of graphene can greatly reduce the interfacial thermal resistance. With the increase of hydrogen atom coverage, the relative interfacial thermal resistance decreases[34]. Luo et al. used non-equilibrium molecular dynamics methods to study the factors affecting the thermal resistance of the interface between graphene and polymer matrix. They proposed that increasing the polymer density, increasing the size of graphene, and forming covalent bonds are conducive to interfacial heat transfer[35].

In this paper, we will study the thermophysical properties of the composite system of modified graphene (MG) and n-octadecane through molecular dynamics simulation and compare the simulation results with the experimental test results.

2. Structure models and simulation methods

Due to the limitation of simulation conditions, the MG model used in the simulation is idealized and simplified. A supercell structure was established to obtain a single-layer graphene sheet containing 160 carbon atoms. Four functional groups are connected at the edges of the graphene sheet near the four corners. A model of the MG is shown in Fig 1. The gray atoms in the picture represent carbon atoms, the white atoms represent hydrogen atoms, and the red atoms are oxygen atoms.

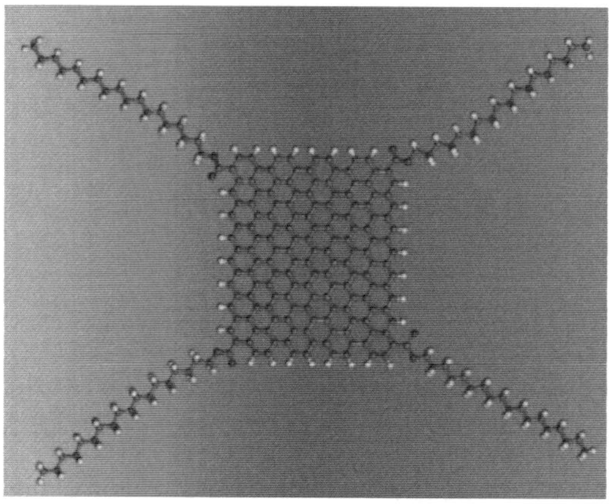

Fig 1. Model of modified graphene.

The Amorphous cell module of Materials Studio software was used to construct the initial model of the MG/n-octadecane composite PCM. In order to reflect the difference in the amount of MG, a total of 6 models have been established, in which the number of MG molecules is 0, 1, 2, 3, 4, 5, respectively. For the convenience of description, these sets of models were labeled as MG0, MG1, MG2, MG3, MG4 and MG5, respectively. At the same time, in order to better simulate the distribution of MG in n-octadecane, the MG was fixed parallel to each other in the center of the simulation box. Fig 2 is the initial model of MG3. The yellow part is n-octadecane molecule and the blue part is MG.

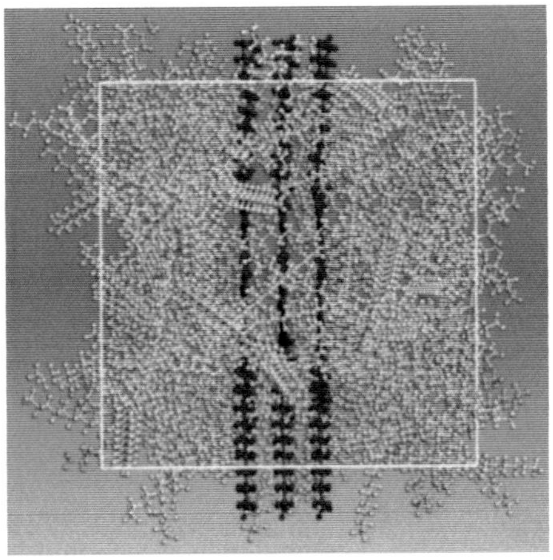

Fig 2. Initial model of MG3.

After the initial model is built, there are some irrationals in the system, so it is necessary to perform a relaxation process and optimize the structure to obtain a system as close to the actual situation as possible. The relaxation process includes energy minimization, annealing, and kinetic equilibrium processing. After the relaxation process is completed, from the energy and temperature curves of the entire process, we can see that the system is basically stable, as shown in Fig 3.

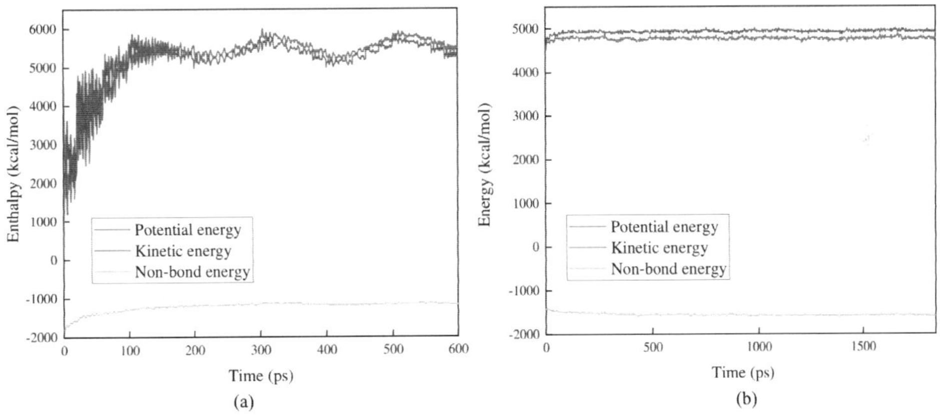

Fig 3. Energy variation of relaxation process (a)Energy variation of annealing process; (b)Energy variation of kinetic equilibrium process.

After the system reached equilibrium, the COMPASS force field was selected to start dynamic calculations at different temperatures between 280K and 330K. Within the set temperature interval, calculations are performed every 5K intervals.

3. Results and discussions

3.1Results of thermal conductivities of composites

The thermal conductivities of six groups of MG composite PCM systems were calculated using the RNEMD method[36], the results are shown in Fig 4. As the temperature increases, the thermal conductivity of each system generally decreases. It can be seen that when the temperature exceeds 300K, the thermal conductivity decreases significantly. Taking MG0 as an example, the phase transition temperature of n-octadecane is 28°C (301K), and the region where the thermal conductivity significantly decreases can be considered as the phase transition occurrence region (295K-305K). The calculated results show that the average thermal conductivity of solid and liquid n-octadecane are 0.222W/(m·K) and 0.198W/(m·K), respectively. It's worth noting that the thermal conductivity increases slightly between 305 and 310K, which is the end of the phase transition. The variation of the thermal conductivity of the other groups of MG PCM systems at different temperatures is basically consistent with MG0. The decrease of the thermal conductivity and the slight increase of the thermal conductivity after the phase transition is related to the phase transition behavior of n-octadecane, which is the result of the effect of the phase transition process on the molecular order and molecular motion in the system.

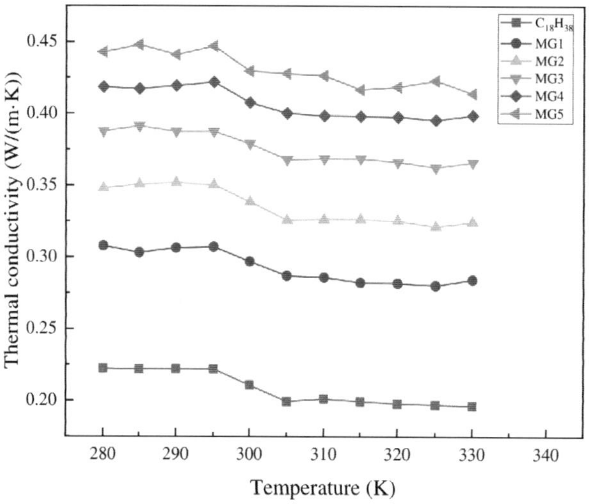

Fig 4. Thermal conductivity of various systems at different temperatures.

The mass fraction of MG in six composite PCM systems was calculated as 0%, 5.40%, 10.80%, 16.69%, 21.54% and 26.88%, respectively. The thermal conductivity calculated at the temperature of 293K is compared with the thermal conductivity measured experimentally, and the results are shown in Fig 5. The experimental and simulation results show that the thermal conductivity of the composite PCM increases with increased MG content. However, the experimentally measured thermal conductivity increases much faster with the increased modified graphene mass fraction than the simulation results. This is because it is difficult to simulate the space heat conducting mesh structure formed by the modified graphene in the n-octadecane system under the existing simulation conditions.

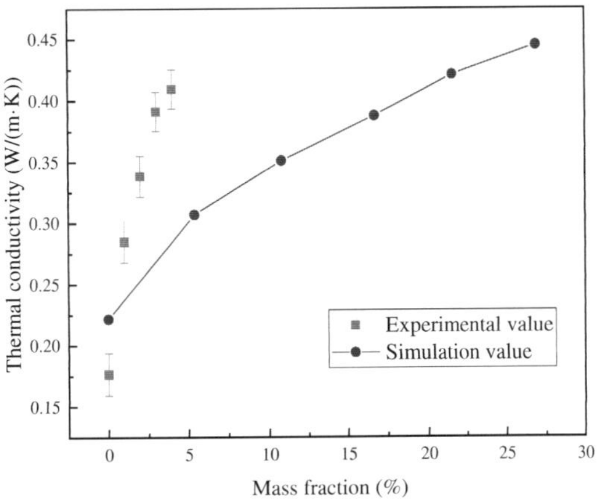

Fig 5. Comparison of experimental and simulation results of thermal conductivity.

3.2Discussions about results of thermal conductivities of composite

During the phase transition, when the solid material absorbs heat and melts into a liquid state, the entropy value of the system increases, and the degree of order decreases. Through molecular dynamics simulation, we can observe that with the change of temperature, the atoms in the system move continuously, and their arrangement also changes, which leads to changes in the molecular arrangement and configuration in the model. We can research the changes of order degree in the composite PCMs system by end-to-end distance and torsion angle of the n-octadecane molecular chain at different temperatures, as shown in Fig 6.

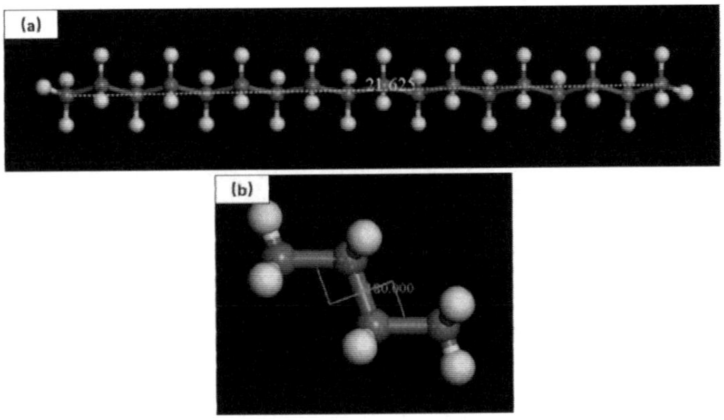

Fig 6. Schematic diagram of end to end distance and torsion angle (a) end to end distance; (b) torsion angle.

Fig 7 shows the end-to-end distance distribution of the n-octadecane PCM at different temperatures. As shown in Fig 7, at 285K, the terminal distance distribution has a highest peak between 17 Å and 20 Å, indicating that at low temperatures, the arrangement of n-octadecane is relatively neat, with a high order degree. As the temperature increases, we can clearly observe that the peaks in the range of 17Å to 20 Å are getting shorter and shorter, and the highest peak has a tendency to shift to the left. The distribution of molecules with end-to-end distance less than 15 Å are more random, with no obvious regularity, means that the degree of disorder of the n-octadecane molecule is getting higher and more random.

Fig 7 is a distribution diagram of torsion angles of n-octadecane molecules at different temperatures. The peak value in the coplanar ±180° region gradually decreases as the temperature increases, and the peak value in the ±160° and ±75° regions gradually appears, indicating that the degree of twisting and bending of n-octadecane molecules increases under the action of heating up, and the consistency and order degree of molecules decrease.

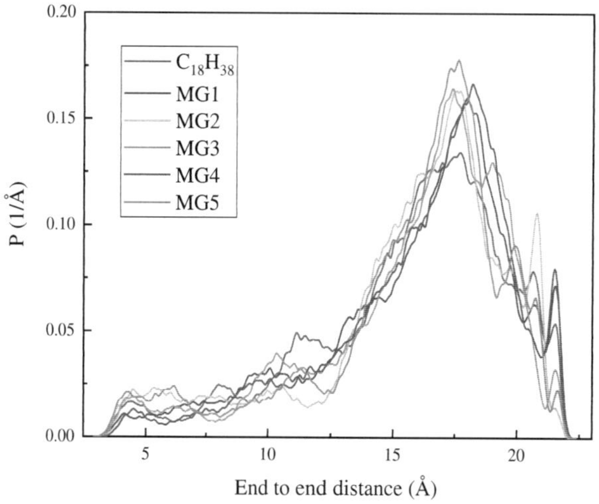

Fig 7. Distribution curve of end to end distance of n-octadecane molecules at different temperatures during heating.

When the temperature exceeds the phase transition point, it can be considered that the ordered arrangement of the solid n-octadecane system and the consistency of the molecules are destroyed. According to the principle of solid heat conduction, there are two types of heat transfer inside solid materials, which are lattice heat conduction and electron heat conduction. For n-octadecane, it mainly depends on lattice conduction. The normal mode energy quantum of the lattice vibration is called a phonon. The heat conduction in the crystal depends on the motion of the phonon. The microscopic expression of its thermal conductivity is shown in formula (1)

$$k = \frac{1}{3} C v \overline{V} l \tag{1}$$

In the formula, c_v represents the heat capacity per unit volume, $\overline{v}$ represents the average group velocity of the phonon, and l represents the mean free path of the phonon.

When the orderness of the n-octadecane PCM decreases due to melting, the mean free path of the phonon also decreases. According to formula (1), the thermal conductivity decreases.

In addition, during the phase transition, molecules in the solid and liquid states coexist in the n-octadecane system. At the interface between the solid and liquid, the molecule's movement is hindered. After the phase transition is completed, the blocked state disappears, thermal conductivity will rise slightly. This explains the increase in thermal conductivity from 305K to 310K in Fig 4.

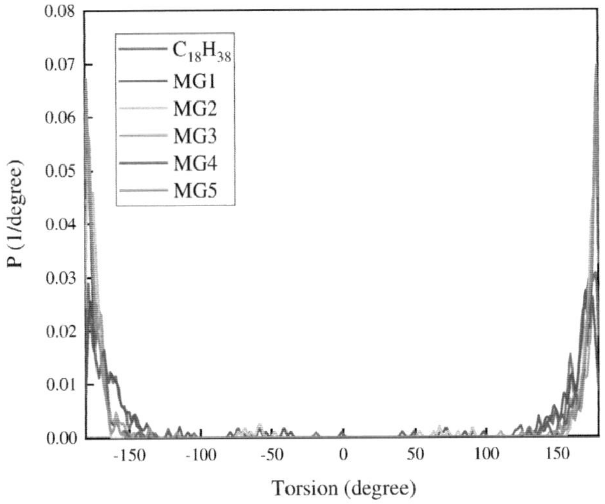

Fig 8. Distribution curve of torsion angle of n-octadecane molecules at different temperatures during heating.

The end-to-end distance and torsion angle distributions of n-octadecane molecules at 325K in pure n-octadecane and MG n-octadecane systems are shown in Fig 9 (a) and (b), respectively. In Fig 9 (a), at 325K, the end-to-end distance of the pure octadecane system is mostly distributed between 15Å and 20Å. As the amount of MG in the system increases, the peaks around 17.5Å become higher, and the peaks in other ranges gradually become shorter. In Fig 9 (b), as the amount of MG increases, the peak around ± 180 ° rises, indicating that the degree of bending of the n-octadecane molecules in these systems are getting lower, until MG5, which exceeds 80% of n-octadecane. The torsion angle of the alkane molecule is not less than ± 170°, showing a high consistency. According to the previous analysis, this is the expression of increasing order of n-octadecane molecules in these systems. In other words, the addition of MG can improve the order degree of molecules in the PCM system.

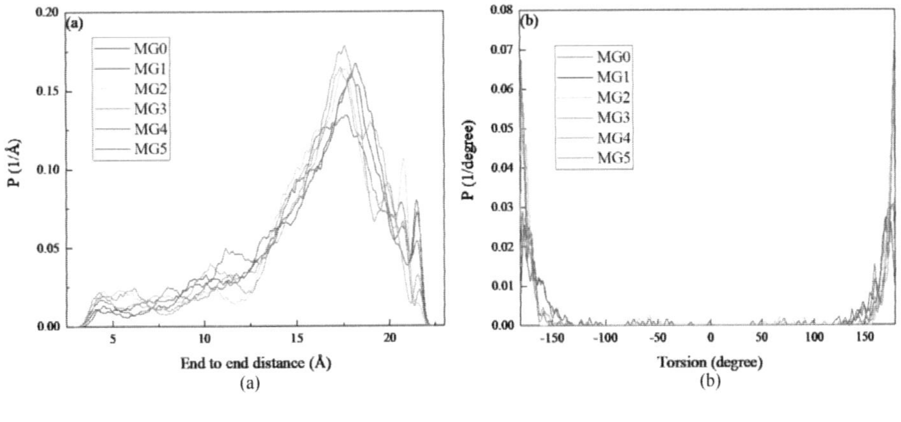

Fig 9. Distribution curve of end to end distance and torsion of n-octadecane molecules in various systems.

3.3 Radial distribution function

The physical meaning of the Radial Distribution Function (RDF) is the ratio of the probability of another particle appearing at a distance "r" relative to the random distribution of a reference particle. The formula is as follows [37]:

$$g(r) = \frac{dN}{\rho 4\pi^2 dr} \tag{2}$$

In the formula, N represents the number of particles, and ρ represents the density of the entire system.

We can use the RDF to reflect the aggregation and binding of molecules in the n-octadecane PCM system. The results are shown in Fig10. There are peaks at 1.125 Å and 1.145 Å, and smaller peaks near 2Å to 3Å. The higher the peak value, the higher the proportion of n-octadecane molecules maintaining the corresponding distance. The positions of the two highest peaks did not change substantially with temperature, indicating that temperature did not significantly affect the aggregation and binding mode of n-octadecane molecules. As the temperature increases, we can observe that all peaks become shorter, which means that the number of molecules that maintain the corresponding distance gradually decreases. This may be due to the higher kinetic energy of n-octadecane at higher temperatures, the greater the relative distance between molecules, and the increased mobility of the molecules in the system. At the same time, the decrease of the number of molecules keeping the same relative distance can also be interpreted as the decrease of the order degree of the n-octadecane system.

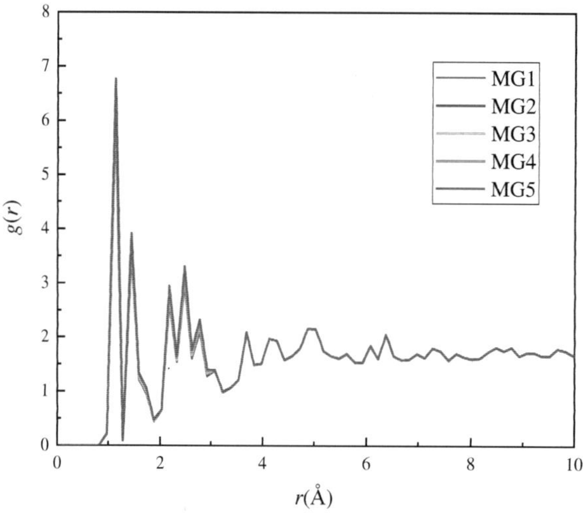

Fig 10.RDF of n-octadecane molecules at different temperatures.

Fig 11 shows the radial distribution function of n-octadecane molecules in MG1, MG2, MG3, MG4 and MG5. As shown in Fig 11, the value of the RDF increases with the increase of the number of MG molecules, but the change is not obvious. This is because only n-octadecane occurs in the phase change process of the composite PCMs, and the MG does not significantly affect the process. Since the MG was fixed in the center of the simulation box, the thermal movement of n-octadecane molecules was limited, and their mobility was reduced, resulting in a slightly higher radial distribution function.

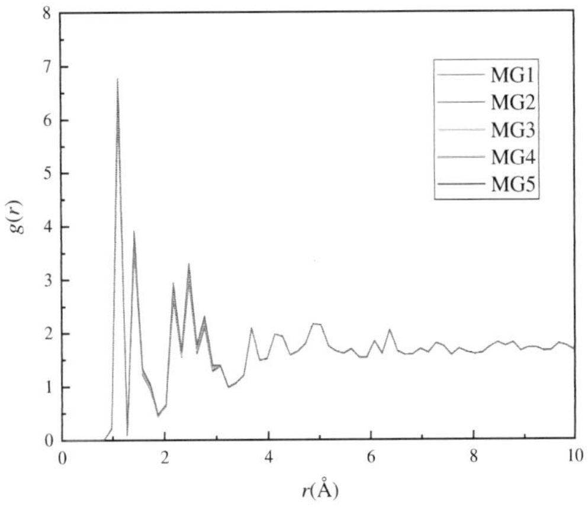

Fig 11.RDF of n-octadecane molecules in each system at 325K.

3.4 Results and discussions of Phase change characteristics

3.4.1 Self-diffusion coefficient

The phase change characteristics of PCMs in molecular dynamics simulation can be studied by the self-diffusion coefficient. The self-diffusion coefficient is a single-valued function of temperature. It is generally used to describe the diffusion behavior of atoms or molecules in molecular dynamics simulation. Its numerical value reflects the strength of the mobility of atoms in the system. If the value of the self-diffusion coefficient changes abruptly, it indicates that the system has undergone a phase change.

Fig 12 is a curve of the mean square displacement (MSD) of the n-octadecane PCM system at different temperatures during the simulated temperature rise. With the development of the simulation, the kinetic energy of molecules in the system gradually increased, and MSD showed an increasing trend. At the same time, the higher the simulated temperature, the higher the kinetic energy obtained by the molecules in the same time, so the MSD value is higher.

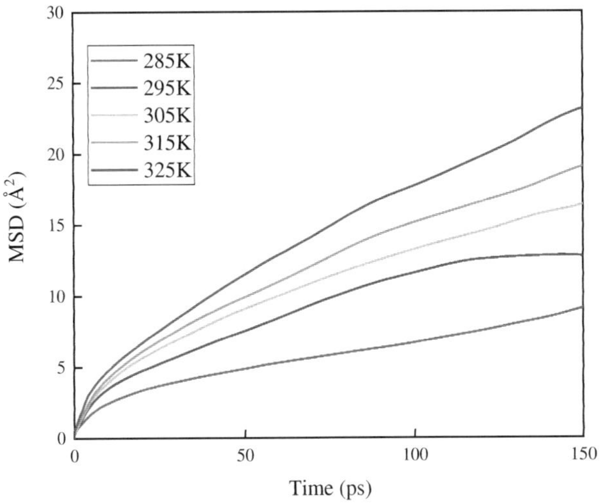

Fig 12. Mean square displacement curve of n-octadecane system at different temperatures during heating.

The slope of n-octadecane molecule MSD at each temperature is calculated. After calculation and processing, the self-diffusion coefficient of n-octadecane PCM systems in the temperature range of 280K to 330K is obtained[38,39], and the result is shown in Fig 13. The value of the self-diffusion coefficient increases with increasing temperature, indicating that the molecular mobility in the n-octadecane system increases with increasing temperature [41]. The self-diffusion coefficient scatter points of two temperature ranges from 280K to 300K and from 305K to 330K were linearly fitted, and two straight lines were obtained. The temperature at the intersection was 302.6K. This point can be regarded as the phase transition temperature of the composite PCM [32]. MG promotes the crystallization of n-octadecane molecules at this temperature[31]. Compared with the experimentally measured phase transition temperature (301.6K) of n-octadecane, the error does not exceed 1%, which proves that the simulation results are reliable.

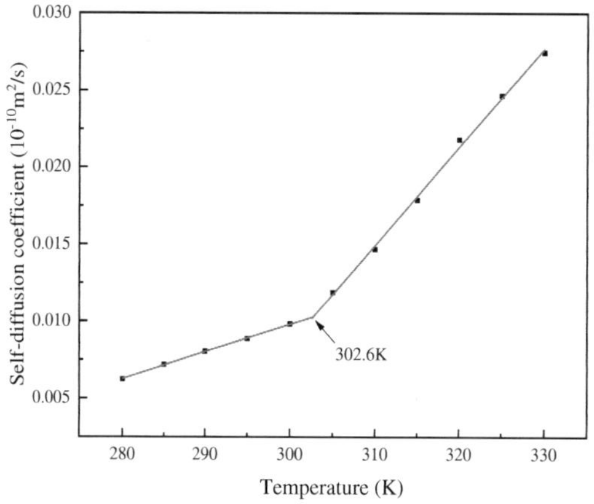

Fig 13. Self-diffusion coefficients of n-octadecane system during heating.

Fig 14 shows the MSD of n-octadecane molecules in each composite PCM system at 325K. The MSD and self-diffusion coefficient decrease as the amount of MG increases. Fig 15 is a scatter diagram of the self-diffusion coefficient of each group of MG PCM systems with the change of temperature. The phase transition temperature can be obtained by fitting the scatter points to a straight line.

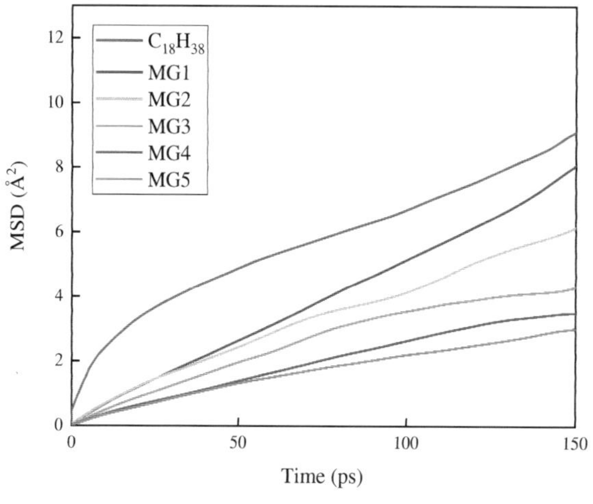

Fig 14.MSD of n-octadecane molecules in each system at 325K.

Fig 15 is a scatter diagram of the self-diffusion coefficient of each group of MG/PCM systems with the change of temperature. The phase transition temperature can be obtained by fitting the scatter points to a straight line. As shown in Fig 15, all the systems follow the rule that the self-diffusion coefficient increases with the increase of temperature. The phase transition temperature decreased slightly with the increase of the content of the MG (Fig 15(b) , and the overall trend slightly increased compared to Fig 15(a)), and the magnitude of the change was small, which was consistent with the experimental conclusion. This may mean that some interactions between the MG and n-octadecane caused such changes.

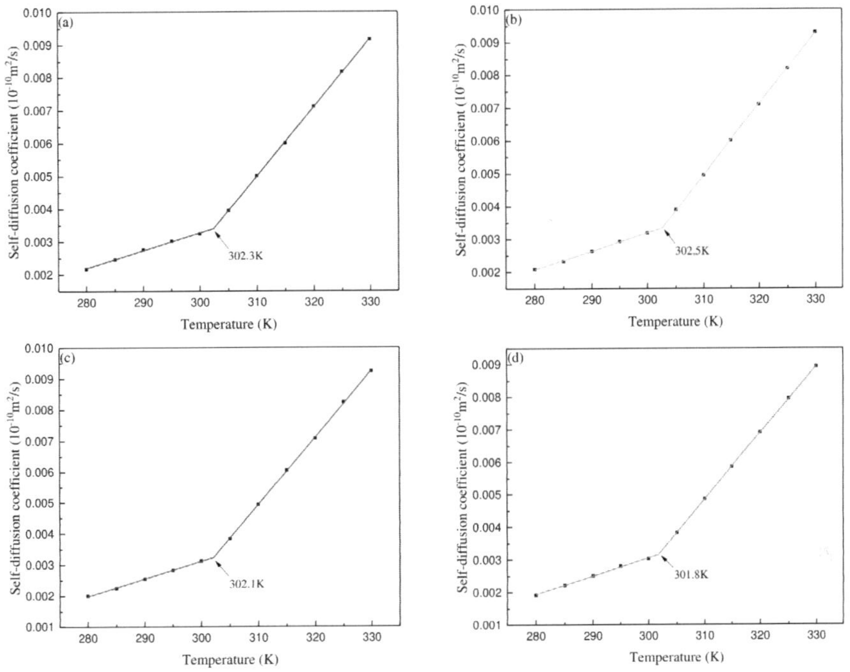

Fig 15. Self-diffusion coefficients of molecules in each system during heating (a)MG1; (b)MG2; (c)MG3; (d)MG4.

3.4.2 Specific heat capacity

In addition to the latent heat of phase change, the specific heat capacity of the PCM affects its heat storage capacity. In molecular dynamics simulations, we calculate the derivative of the total potential energy of the system with respect to temperature, and obtain the constant heat capacity C_V

$$C_V = \frac{dE(T)}{dT} \tag{3}$$

The potential energy curve of the n-octadecane PCM system during the temperature rise process is drawn. As shown in Fig 16, the potential energy of the system increases with increasing temperature, which approximates a linear increase.

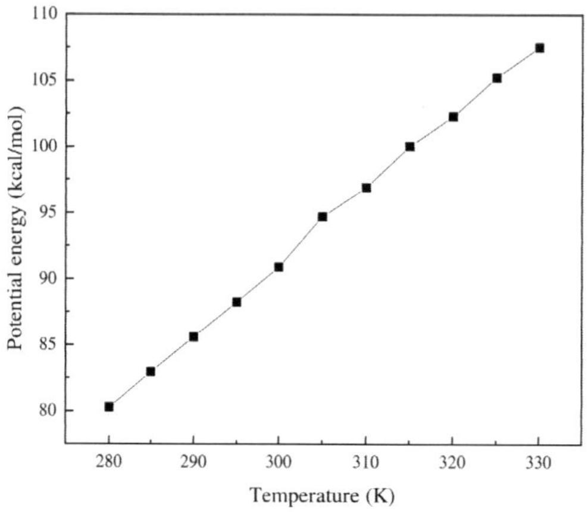

Fig 16. Potential energy curve of n-octadecane system during heating.

According to equation (3), take the first derivative of the curve in Fig 16, and the change curve of the specific heat capacity of the n-octadecane PCM system with temperature is obtained, and the result is shown in Fig 17.

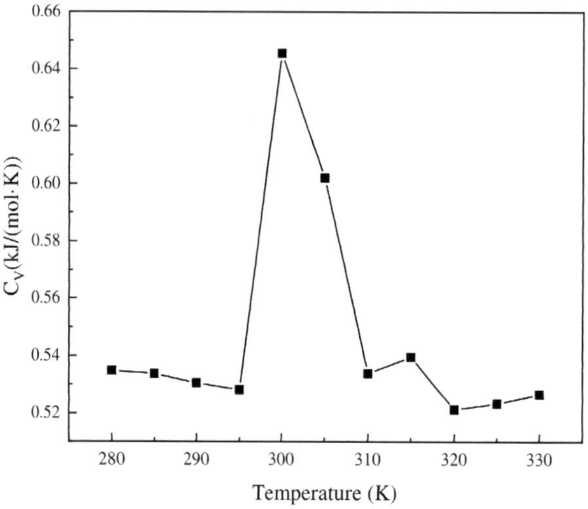

Fig 17. Specific heat capacity variation of n-octadecane system with temperature.

When the PCM does not undergo a phase change, the heat capacity is a single value function of temperature. If the phase change occurs, the value of the heat capacity will change greatly, but the temperature will not change. Therefore, if the curve of specific heat capacity with respect to temperature changes abruptly, it indicates that phase change occurs in the PCM. As shown in Fig 17, the specific heat capacity at 300K increased significantly, which is close to the experimental value of the phase transition temperature of n-octadecane (301.6 K), which proves that the simulation results are reliable. Although we cannot obtain the latent heat value of the n-octadecane system through simulation, we can see that the specific heat capacity value near the phase transition temperature jumps significantly due to the latent heat, which is obviously conducive to thermal energy storage. The experimental measured value is compared with the simulated specific heat capacity value. When the temperature is 293K, we measured the experimental value as 1.658J/(g·K) and the simulated value as 2.083J/(g·K). The difference between the two values may come from the COMPASS force field, which is the most suitable force field for describing condensed-phase materials, but it is also difficult to accurately describe the accurate thermodynamic behavior of complex organic compound molecules [40]. Nevertheless, the simulation results can still be used as a reference for studying the thermophysical properties of PCMs.

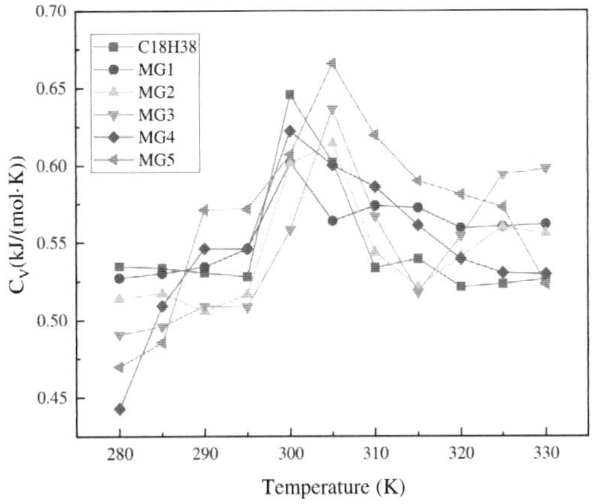

Fig 18. Specific heat capacity variation of each system with temperature.

The specific heat capacity values of MG0, MG1, MG2, MG3, MG4, and MG5 are compared, and the results are shown in Fig 18. Each curve rises sharply at 300K or 305K, which is the effect of the latent heat of the composite PCM. The addition of MG does not significantly affect the specific heat capacity change law of PCMs.

4. Conclusion

In this paper, the molecular dynamics simulation was used to calculate the n-octadecane and MG composite PCM system, and the results were compared with the experimental results. The following conclusions are obtained:

(1) Comparing the simulation results of the composite phase change system with the n-octadecane system, it is found that the trend of thermal conductivity with temperature changes is the same as the experimental results, and the value of thermal conductivity increases with the increase in the content of MG.

(2) Analysis of the parameters such as the end-to-end distance, torsion angle, RDF, and self-diffusion coefficient of each system shows that the reason for the increase in thermal conductivity of the composite PCMs is that the interaction between MG and n-octadecane caused an increase in the order of n-octadecane, leading to an increase in the thermal conductivity of n-octadecane itself. This is the main factor affecting the thermal conductivity of composite PCMs in molecular dynamics simulation.

(3) Simulation results show that the phase transition temperature of PCM decreases with the addition of MG. The difference between the calculated value of the phase transition temperature and the experimental value is less than 1%.

Acknowledgments
The work is supported by the National Natural Science Foundation of China (No.51606017), the Youth Talent Support Program of Chongqing (CQYC2021059206), Chongqing Science Fund for Distinguished Young Scholars (cstc2021jcyj-jqX0015), and the Fundamental Research Funds for the Central Universities (No.106112016CDJXY145502).

References

[1] Zhang P, Ma Z W and Wang R Z 2010 *Renew. Sustain. Energy Rev.* **14** 598–614
[2] Zhang Q, Wang H, Ling Z, Fang X and Zhang Z 2015 *Sol. Energy Mater. Sol. Cells* **140** 158–66
[3] Zhang Z, Zhang N, Peng J, Fang X, Gao X and Fang Y 2012 *Appl. Energy* **91** 426–31
[4] Lin Y, Jia Y, Alva G and Fang G 2018 *Renew. Sustain. Energy Rev.* **82** 2730–42
[5] Halime Paksoy 2007 *Thermal Energy Storage for Sustainable Energy Consumption:Fundamentals,Case Studies and Design*
[6] Karaman S, Karaipekli A, Sarı A and Biçer A 2011 *Sol. Energy Mater. Sol. Cells* **95** 1647–53
[7] Dao T D and Jeong H M 2015 *Sol. Energy Mater. Sol. Cells* **137** 227–34
[8] Mohamed N H, Soliman F S, El Maghraby H and Moustfa Y M 2017 *Renew. Sustain. Energy Rev.* **70** 1052–8
[9] Chandel S S and Agarwal T 2017 *Renew. Sustain. Energy Rev.* **67** 581–96
[10] Pandey A K, Hossain M S, Tyagi V V, Abd Rahim N, Selvaraj J A /L. and Sari A 2018 *Renew. Sustain. Energy Rev.* **82** 281–323
[11] Wu Z, Qin M and Zhang M 2018 *Energy Build.* **174** 254–61
[12] Yang J, Tang L-S, Bao R-Y, Bai L, Liu Z-Y, Yang W, Xie B-H and Yang M-B 2017 *Chem. Eng. J.* **315** 481–90
[13] Zhang M, Medina M A and King J B 2005 *Int. J. Energy Res.* **29** 795–809
[14] Li Y, Du Y, Xu T, Wu H, Zhou X, Ling Z and Zhang Z 2018 *Appl. Therm. Eng.* **131** 766–78
[15] Jaguemont J, Omar N, Van den Bossche P and Mierlo J 2018 *Appl. Therm. Eng.* **132** 308–20
[16] Setoh G, Tan F L and Fok S C 2010 *Int. Commun. Heat Mass Transf.* **37** 1403–10
[17] Ibrahim N I, Al-Sulaiman F A, Rahman S, Yilbas B S and Sahin A Z 2017 *Renew. Sustain. Energy Rev.* **74** 26–50
[18] Fan L-W, Fang X, Wang X, Zeng Y, Xiao Y-Q, Yu Z-T, Xu X, Hu Y-C and Cen K-F 2013 *Appl. Energy* **110** 163–72
[19] Zhang Z and Fang X 2006 *Energy Convers. Manag.* **47** 303–10
[20] Jin Y, Wan Q and Ding Y 2015 *Procedia Eng.* **102** 1877–84
[21] Anon Carbon nanoadditives to enhance latent energy storage of phase change materials: Journal of Applied Physics: Vol 103, No 9
[22] Cui Y, Liu C, Hu S and Yu X 2011 *Sol. Energy Mater. Sol. Cells* **95** 1208–12
[23] Yuan K, Shi J, Aftab W, Qin M, Usman A, Zhou F, Lv Y, Gao S and Zou R 2020 *Adv. Funct. Mater.* **30** 1904228
[24] Cai Di and Li Jing 2019 *Acta Phys. Sin.* **68** 18–26
[25] Yu W, Xie H, Li Y, Chen L and Wang Q 2011 *Colloids Surf. Physicochem. Eng. Asp.* **380** 1–5

[26] N.-M. Barkoula, B. Alcock, and N.O. Cabrera 2008 *Polym. Compos.* **380** 101–13
[27] Park M, Lee T and Kim B-S 2013 *Nanoscale* **5** 12255–60
[28] Di C J L 2020 *J. Chem. Ind. Eng.* **71** 4826–35
[29] Sosso G C, Miceli G, Caravati S, Giberti F, Behler J and Bernasconi M 2013 *J. Phys. Chem. Lett.* **4** 4241–6
[30] Yin Kai-liang, Duan-Jun, and Xia Qing 2001 *Acta Phys.-Chim. Sin.* **20** 302–5
[31] Babaei H, Keblinski P and Khodadadi J M 2013 *Int. J. Heat Mass Transf.* **58** 209–16
[32] Rao Z, Wang S, Wu M, Zhang Y and Li F 2012 *Energy Convers. Manag.* **64** 152–6
[33] Zabihi Z and Araghi H 2016 *Phys. Lett. A* **380** 3828–31
[34] Wang Y, Yang C, Cheng Y and Zhang Y 2015 *RSC Adv.* **5** 82638–44
[35] Tengfei L and John R. Lloyd 2012 *Adv. Funct. Mater.* **22** 2495–502
[36] Müller-Plathe F 1997 *J. Chem. Phys.* **106** 6082–5
[37] Zhou G, Zhao T, Wan J, Liu C, Liu W and Wang R 2015 *Carbohydr. Res.* **401** 89–95
[38] Liu X and Rao Z 2019 *J. Energy Inst.* **92** 161–76
[39] Chang T, Hai J, Zhou J, and Ling-Hong L 2009 *Acta Phys.-Chim. Sin.* **25** 1373–8
[40] Wang Y, Chen Z and Ling X 2016 *Appl. Therm. Eng.* **98** 835–40

Color-changing Fluorescent Carbon Dots Synthesized by Simple Hydrothermal Method for Anti-counterfeiting Pattern Study

Qidan Chen[1,2 *], Youyi Xu[1,2], Maolin Sun[1], Xinyao Zang[1,2], and Jinglan Qiu[1]

[1] School of Applied Chemistry and Materials, Zhuhai College of Science and Technology, Zhuhai 519041, China

[2] Zhuhai Institute of Supramolecular Materials, Southern Institute, Jilin University, Zhuhai 519041, China

*qidanchen@jlu.edu.cn

Abstract. A simple hydrothermal method was employed to synthesize the carbon dots with dopamine hydrochloride used as the mainly carbon source and o-phenylenediamine as the nitrogen source. In the study, the fluorescent carbon dots which have good water solubility and alcohol solubility showed visible pink under daylight and orange under UV after purification by column chromatography, and dichloromethane: methanol (v: v =20:1) was selected as the separation solvents. Meanwhile, the pink fluorescent carbon dots (λmax =595 nm under 577 nm excitation) was observed that it can be changed to bright yellow (λmax =530 nm under 442 nm excitation), the conditions of different stocked solvents and time were explored in detail. The IR spectra for the pink carbon dots indicated its color-changing property due to the unsatisfaction group of its chemical structure. For UV and fluorescence characterization, the quantum yield of pink carbon dots is ~11.0%. In addition, the application of color-changing carbon dots as anti-counterfeiting ink for high-quality anti-counterfeit patterns on paper was investigated, and the preliminary results showed good potential of the prepared carbon dots as effective and eco-friendly anti-counterfeiting materials.

Keywords: Color-changing, Fluorescent carbon dots, Hydrothermal method, Anti-counterfeiting

1. Introduction

In recent years, the development of science and technologies has provided more convenient materials and diverse products in daily life. However, the identification of the products from the counterfeiters for consumers is more and more important. In addition, a simple products label with above function and information such as expiration date can provide more protection for products and convenience for consumers. Therefore, it is necessary to explore more efficient and eco-friendly anti-counterfeiting materials. In this work, carbon dots as a novel anti-counterfeiting material were preliminary studied for the conditions of preparation, purification, and application as anti-counterfeiting ink. As a new type of fluorescent carbon nanomaterial, carbon dots were selected due to its excellent optical properties, high biocompatibility, low cytotoxicity, and less harmful to the environment. Carbon dots also have the advantages of its green source materials, simple preparation process and low-cost, which have great potential for fluorescent anti-counterfeiting application.

Carbon-based materials such as carbon nanotubes, graphene, carbonized polymers were widely investigated as fluorescent nanomaterials have been paid of great attention in optical application due to the simple synthesis and eco-friendly preparation method [1]. The original sources for carbon dots preparation can be various by chemical or even biomass [2]. For example, Yang group used citric acid as carbon source and urea as nitrogen source to synthesize fluorescent carbon dots [3]. And Zulfajri group used oyster mushroom as carbon source to synthesize blue fluorescent carbon dots with specific response to nitroaromatic hydrocarbon [4]. Nowadays, the excellent optical properties as well as low toxicity and biocompatibility of carbon-based material have been widely applied in many fields.

Chemistry anti-counterfeiting methods basically includes natural light, ultraviolet photochromic and thermochromic, which are mainly applied by form of ink printing for product trademarks and packaging patterns. The anti-counterfeiting technology developing is more simple, low cost, good concealability, color-changing feasibility, reproducibility, and the fluorescent carbon dots material which excellent optical properties probably meet the development requirements are investigated by researchers as novel anti-counterfeiting material [5] [6].

In this paper, carbon dots were synthesized by hydrothermal method with dopamine hydrochloride as the carbon source and o-phenylenediamine as the modified nitrogen source, a series of fluorescent carbon nanomaterials with visible pink while orange under UV and which color can be changed to bright yellow under certain conditions and studied for the anti-counterfeiting application by testing of anti-counterfeit patterning on paper.

2. Materials and methods

Fluorescent carbon dots of pink (PB-CDs) were prepared by hydrothermal method under the temperature of 180°C for 16 hr using dopamine hydrochloride (Energy Chemical, Shanghai, China) as carbon source and o-phenylenediamine (Energy Chemical, Shanghai, China) as nitrogen source in mixture of water and ethanol. And the ratio of dichloromethane and methanol was 18:1 as the developing agent for column chromatography purification. The UV, fluorescence, and IR spectra were obtained by using Shimadzu Spectrophotometer mode of UV-2450, RF-5301pc, and IR Prestige-21, JEM-2100F Transmission Electron Microscope (Nippon Electron Co. Ltd). Calculated by measurement method [7], the fluorescence quantum yield of PB-CDs is 11.0 %. Preparing 2 mL of the above PB-CDs stock ethanol solution in glass sample vials and dissolved by 8 mL different solvents of petroleum ether, cyclohexane, dichloromethane, ethyl acetate, n-propanol, ethanol, and methanol separately. The written ink was prepared by 1 g CDs stock solution and 1.5g glycerin, the G-tip of the dipping pen (Lingcat, China), and filter paper（Whatman, GE HealthCare, US）without fluorescent background under the excitation were used for the pattern test.

3. Results and discussions

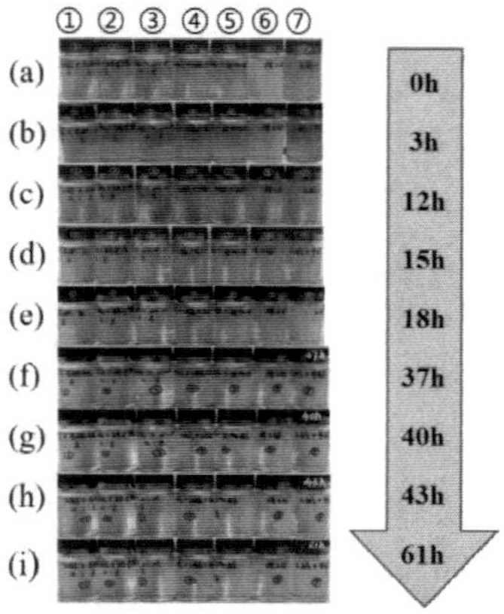

Figure 1. Color-changing of PB-CDs under different storage solvents and time conditions
Storage solvents conditions: ①Ethanol + petroleum ether , ②ethanol + cyclohexane, ③ethanol + dichloromethane, ④ethanol + ethyl acetate, ⑤ethanol + n-propanol, ⑥ethanol + ethanol, ⑦ ethanol + methanol

As Figure 1 shown, the seven samples in different solvents conditions, from ① to ⑦, only the ③ ethanol + dichloromethane sample showed discoloration when it was just configured, while the ④ ethanol + ethyl acetate sample showed orange color in solution due to the color difference of carbon dots in different solvents. After 3 hours of observation, the ③ ethanol + dichloromethane sample showed more obvious discoloration, while the other samples showed little change. After 15 hours, the color of the ethanol + dichloromethane sample had completely turned into yellow, and there was no significant change then. The other samples gradually changed from pink to yellow after 37 hours, and the color was nearly stable after 43 hours, and there was no significant change. The color change of the ③ ethanol + dichloromethane sample was the fastest. The other samples have similar speeds, while the ① ethanol + petroleum ether sample has the slowest rate of discoloration. Within 48 hours, all samples of PB-CDs have changed color to bright yellow (YB-CDs).

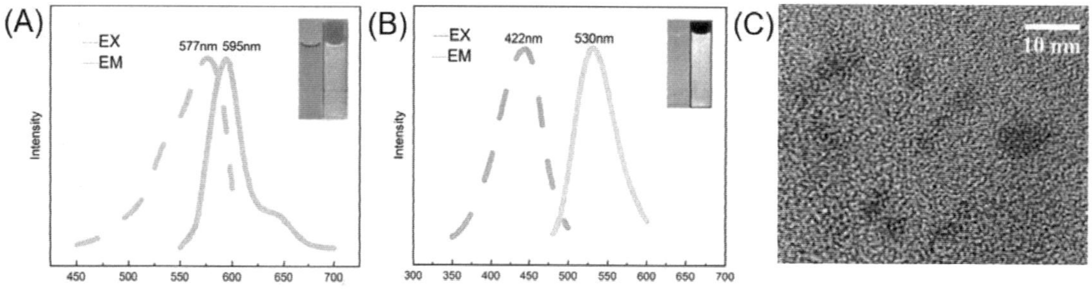

Figure 2. (A) Fluorescence spectra of PB-CDs ethanol solution with inset picture under daylight (left) and UV light (right), (B) YB-CDs ethanol solution with inset picture under daylight (left) and UV light (right), and (C)TEM images of the CDs

Figure 2 shown that the emission wavelength of PB-CDs is 595 nm under excitation at 577 nm, and the emission wavelength of YB-CDs is 530 nm under 442 nm excitation. Both excitation wavelength and emission wavelength were blueshift after the color changed of the material. The carbon dots were also observed by TEM. The TEM image of YB-CDs shown in Figure 2(C) giving direct observation of the carbon dots materials, which are uniformly distributed with the sizes ranged at 4-8 nm.

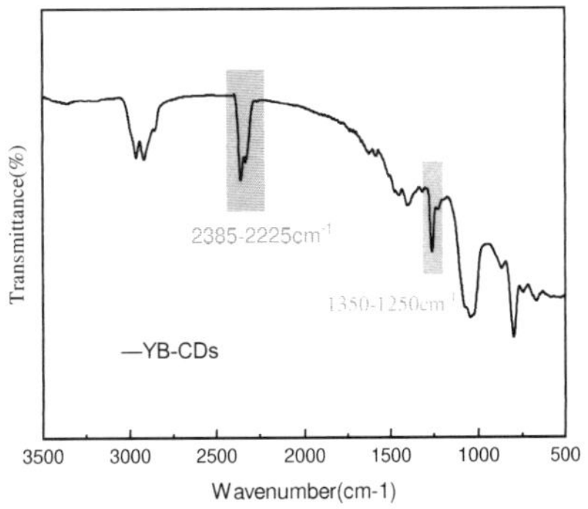

Figure 3. IR spectra of PB-CDs

As Figure 3 shown, the peak in the range of 3000-2850 cm^{-1} is formed by the stretching vibration of the C-H bond. The peak at 1300-1250 cm^{-1} is formed by $-NO_2$ stretching vibration. In addition, the infrared absorption peak at 2385-2225 cm^{-1} indicated that there is an unsaturated cumulative double bond or triple bond structure, which verified that the PB-CDs are easy to be oxidized and color-changed.

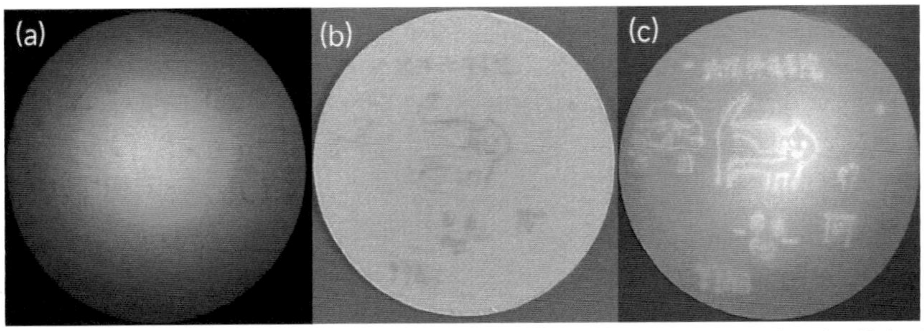

Figure 4. (a) PB-CDs written pattern (b) YB-CDs written pattern under daylight (after color-changing) (c) YB-CDs written pattern under UV (after color-changing)

The ink made of carbon dots with glycerin additive written on filter paper was studied for writing test. The results showed that when the glycerin additive is 12% as mass fraction in the PB-CDs ethanol solution, the handwriting effect is optimized that the ink was well adsorbed in the tip of the dip pen, and very smooth during the writing process which can meet the high-level of brand for 6d neutral ink. As Figure 4 shown, the fluorescent carbon dots ink pattern appears pink under daylight, and a bright fluorescent yellow pattern was observed under UV light (365 nm) after the fluorescent carbon dot ink color-changing.

4. Conclusions

In the work, fluorescent carbon dots applied as anti-counterfeiting material were prepared by simple hydrothermal method and purified by column chromatography. The CDs were characterized by TEM, UV, fluorescence, and IR spectra. In addition, the CDs ink was preliminary investigated for handwriting patterns for anti-counterfeiting application, the investigation may provide a potential new, low toxic, and efficient fluorescent ink materials for the field of anti-counterfeiting.

5. References

[1] Tao S, Feng T, Zheng C, Zhu S and Yang B 2019 Carbonized polymer dots: a brand new perspective to recognize luminescent carbon-based nanomaterials *J. Phys. Chem. Lett.* 10(17): 5182-5188

[2] Wareing T C, Gentile P and Phan A N 2021 Biomass-based carbon dots: current development and future perspectives *ACS nano.* 15(10): 15471-15501

[3] Zhu S, Zhao X, Song Y, Lu S and Yang B 2016 Beyond bottom-up carbon nanodots: Citric-acid derived organic molecules *Nano Today* 11(2): 128-132

[4] Zulfajri M, Rasool A and Huang G G 2020 A fluorescent sensor based on oyster mushroom carbon dots for sensing nitroarenes in aqueous solutions *New J. Chem.* 44 (25): 10525-10535

[5] Lu D, Lu K, Wen H T, Wei Z, Bianco A, Wang G G and Zhang H Y 2023 Synthesis of visible light excitable carbon dot phosphor-Al_2O_3 hybrids for anti-counterfeiting and information encryption *Small* 2207046-2207054

[6] Yu X, Zhang H and Yu J 2021 Luminescence anti-counterfeiting: from elementary to advanced *Aggregate* 2(1): 20-34

[7] Birks J B 1976 Fluorescence quantum yield measurements *J. Res. Natl. Bur. Stand., Sect. A* 80(3): 389-399

Acknowledgements

Supported by Guangdong Special Foundation for the Major Program of Higher Education (2020ZDZX2036), Zhuhai Basic and Applied Basic Research Foundation (ZH22017003200015PWC), Guangdong Provincial Undergraduate Training Program for Innovation and Entrepreneurship (S202313684036), the Three Levels of Talent Project of Zhuhai College of Science and Technology.

Nickel Precipitation on the Surface of 15-15Ti Stainless Steel with B_4C Exposed to Sodium

Bo Qin, Bin Long*

China Institute of Atomic Energy, Xinzhen, Fangshan District, Beijing, China

E-mail: binlong@ciae.ac.cn

Abstract. In order to evaluate the application performance of B_4C pellets as the neutron absorbing material for fast reactor, compatibility experiments were carried out at elevated temperatures for 15-15Ti stainless steel-sodium-B_4C ternary system. Experiments results show that the nickel (Ni) precipitates on the inner surface of the 15-15Ti stainless steel (SS) tube. The Ni precipitates on the surface as island distribution, and with the increase of experiment time, the precipitations grow up. Experiment analysis show that the Ni precipitation process is caused by boronation of the Fe and Cr elements of the 15-15Ti SS tube.

1. Introduction

Sodium-cooled fast reactor (SFR) is the one of the reactors for the fourth-generation advanced nuclear energy system [1]. As the key part of the SFR core system, the control rod element consists of stainless-steel cladding, sodium and B_4C neutron absorber. The cladding materials of control rod is advance austenitic steel 15-15Ti [2]. The compatibility of the 15-15Ti\Na\B_4C ternary system is the key factor for the lifetime evaluation of the cladding materials [3-4].

For the 15-15Ti\Na\$B_4C$ ternary system, the element migration and inner surface corrosion behaviour research is limited [5-6]. The interaction of B and C, and the corrosion behaviour of 15-15Ti cladding tube is unclear. Meanwhile, Ni is key alloy element for stainless steel to sustain the FCC crystalline structure and Ni migration is unknown for 15-15Ti\Na\$B_4C$ ternary system. Hence, it is difficult to evaluate the cladding lifetime using the available experimental data. Therefore, to reveal the Ni migration and transfer behaviour, and to evaluate the application performance of B_4C pellets as the neutron absorbing material for fast reactor, compatibility experiments were carried out at elevated temperatures for the 15-15Ti\Na\$B_4C$ ternary system.

2. Experimental

2.1. Experimental apparatus

In the ternary system, the sodium fills the gap between inner surface and B_4C pellets, and the sodium is purified by a hot trap device, and the Na is located into 15-15Ti cladding tube using an Ar covered glove box. The compatibility experiment is carried out with a sealed high temperature furnace wit PID control system.

2.2. Experimental materials

In this experiment, the control rod cladding material is 15-15Ti SS. The out diameter, inner diameter and length of the cladding tube are 19.2mm, 18.2mm and 200mm, respectively. Table 1 shows the chemical elements of the cladding materials 15-15Ti. The Na is nucleal grad with an oxygen and water content less than 10 ppm. The control rod neutron absorber is manufactured by B_4C powder with natural B abundance. The B_4C pellet is cylindrical with an out diameter and height of 12mm and 15mm, respectively.

Table 1. Chemical concentration of 15-15Ti stainless steel (wt.%)

Element	C	Si	Mn	P	S	Ni	Cr	Mo	Ti	Fe
Content	0.08	<0.45	2.0	<0.035	<0.03	15	15	2.0	0.7	Bal.

2.3. Experimental procedures

Fig.1a shows the experimental assembly of the 15-15Ti\Na\B_4C. In the glove box, two B4C pellets were capsuled into the 15-15Ti SS tube, and then about 15g Na is injected into the 15-15Ti tube. The gap between cladding tube and B4C is filled up by the injected Na. Then the top plug is pressed into 15-15Ti SS tube, and the plug is then welded with the 15-15Ti SS tube (Fig.1b).

Figure 1. Sample assembly schematic diagram and the welded experimental sample capsule.

The compatibility tests were carried out in static high temperature furnace (Fig.1). The compatibility tests were conducted at 600 and 650℃ for 1000 h, 3000 h, 6000 h and 10000h, respectively. After the compatibility tests, the test 15-15Ti tube and B4C were cleaned in alcohol at room temperature to remove the residual sodium on the surface of the specimens. Next, the specimens were also washed by high purity water in ultrasonic cleaning machine. Then the macroscopic morphology of specimens was observed using an optical camera. 15-15Ti cladding tube cross section specimens were prepared by mechanical grinding and polishing. A scanning electron microscope (SEM) with energy dispersion X-ray (EDX) was used to characterize surface morphology and phase transformation on the inner surface of 15-15Ti tube.

3. Results and discussion

Fig.2 depicts the inner surface morphology change of 15-15Ti cladding tube after 1000h and3000h at 600 and 650℃. As shown in Fig.2, the original scratches from machining disappeared, and new phases formed on the surface. In addition, a new precipitation phase was formed on the surface, with an island-shaped distribution. With the increase of the experiment temperature, the growth of precipitates become fast, and with the increment of experiments time, the precipitation phase on the surface become denser.

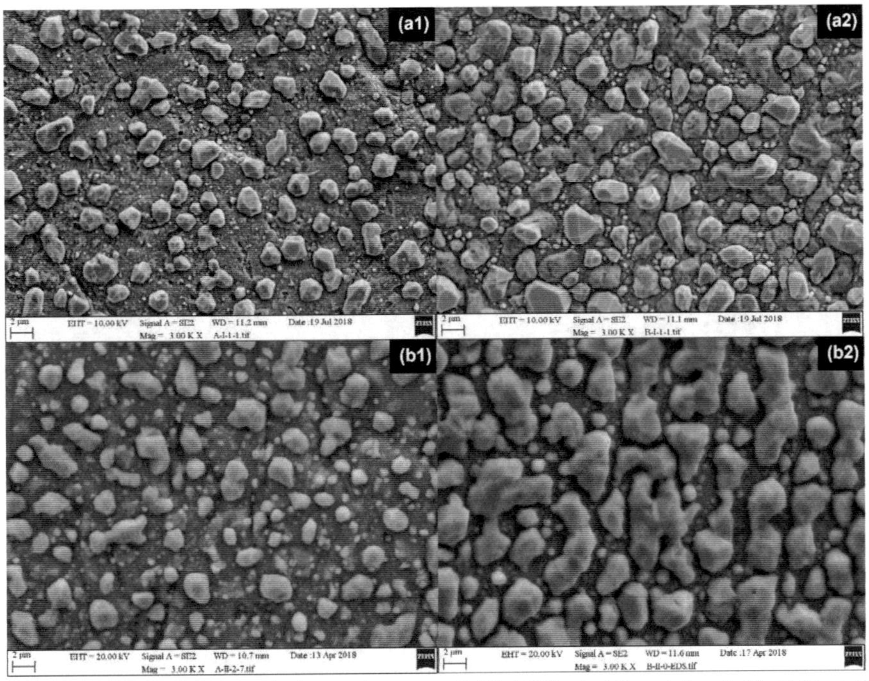

Figure 2. the inner surface morphology changes of 15-15Ti cladding tube. (a1), (b1) at 600℃and (a2), (b2) at 650℃for 1000h and 3000h, respectively.

To illustrate the precipitation behaviour, an EDS element analysis was performed to show the element distribution of the corroded surface. As shown in Fig.3 the main component is Ni. The Ni precipitates from the matrix or corroded layer.

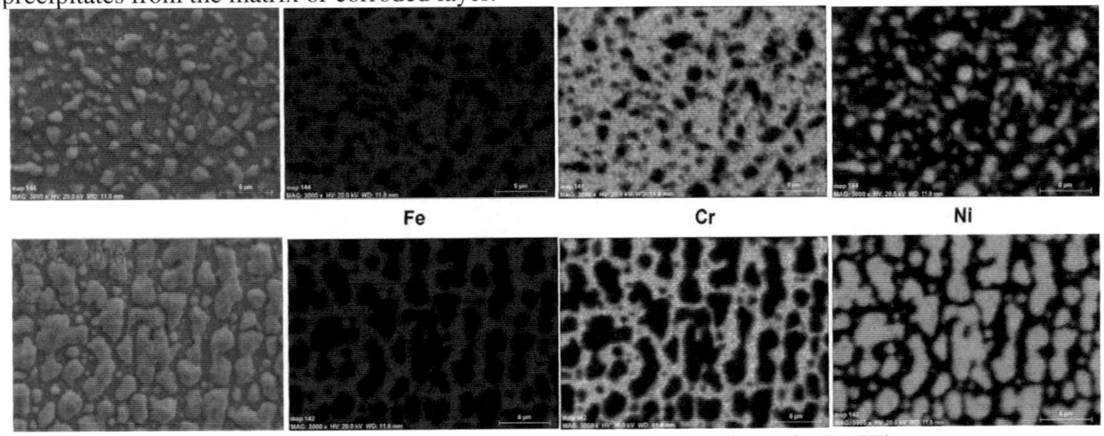

Figure 3. Elements distribution of corroded surface of 15-15Ti

The formation of new precipitation phase on the surface of 15-15Ti cladding tube indicates that inner surface of 15-15Ti cladding was corroded for the 15-15Ti/Na/B_4C ternary system. To illustrate mechanism of the Ni precipitation, the cross section of corroded 15-15Ti was examined through optical microscopy. Fig 4 shows the corroded 15-15Ti section morphology, indicating that a corrosion layer was formed on the surface of the 15-15Ti tube. Meanwhile, an XRD scanning was performed to characterize the corrosion layer phase structure. The XRD spectrum of corroded inner surface of 15-15Ti cladding is shown in Fig.5. The XRD scanning results indicate that the main components of corrosion layer are boride of Fe and Cr. The thickness of the corrosion layer is 6.07 and 11.18μm at 600°C for 1000h and 3000h, 12.87 and 20.89μm at 650 °C for 1000h and 3000h experimental time, respectively.

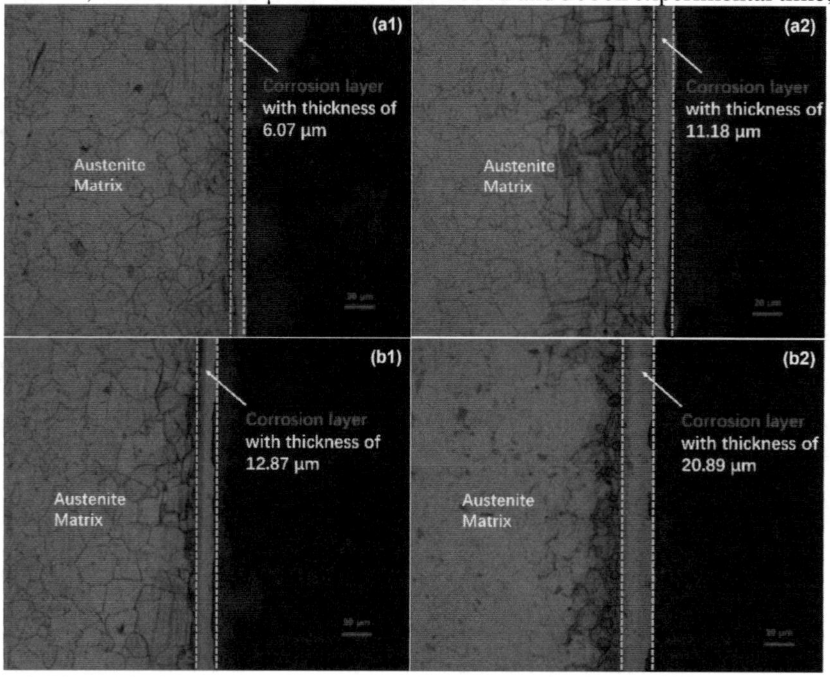

Figure 4. The optical morphology of cross section of 15-15Ti cladding tube inner surface. (a1) and (a2) for 1000h at 600 and 650°C, (b1) and (b2) for 3000h at 600 and 650°C

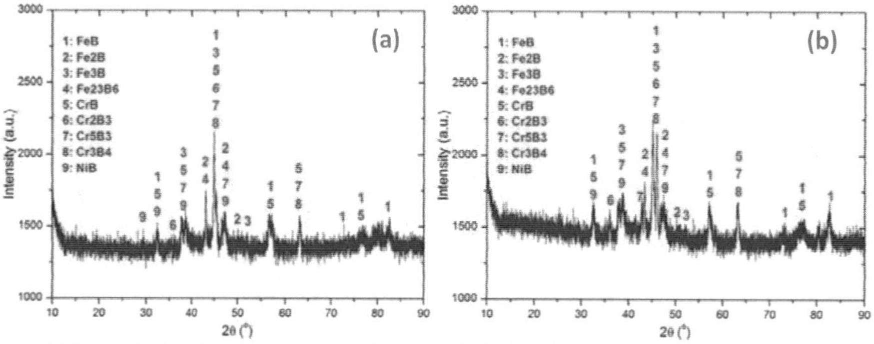

Figure 5. XRD spectrums of the 15-15Ti cladding tube corrosion layer. 3000h experiment at (a) 600 and (b) 650°C

In order to elucidate the correlation between the Ni precipitation and section corroded layer, the EDS mapping of corrosion section of 15-15Ti is performed. Fig 6 shows the elements mapping distribution of cross section of corrosion layer. The EDS mapping results indicate that the Ni concentration in corrosion layer is less than that in austenite matrix, and the corrosion layer mainly contains B, Fe and Cr which is consistent with XRD analysis results.

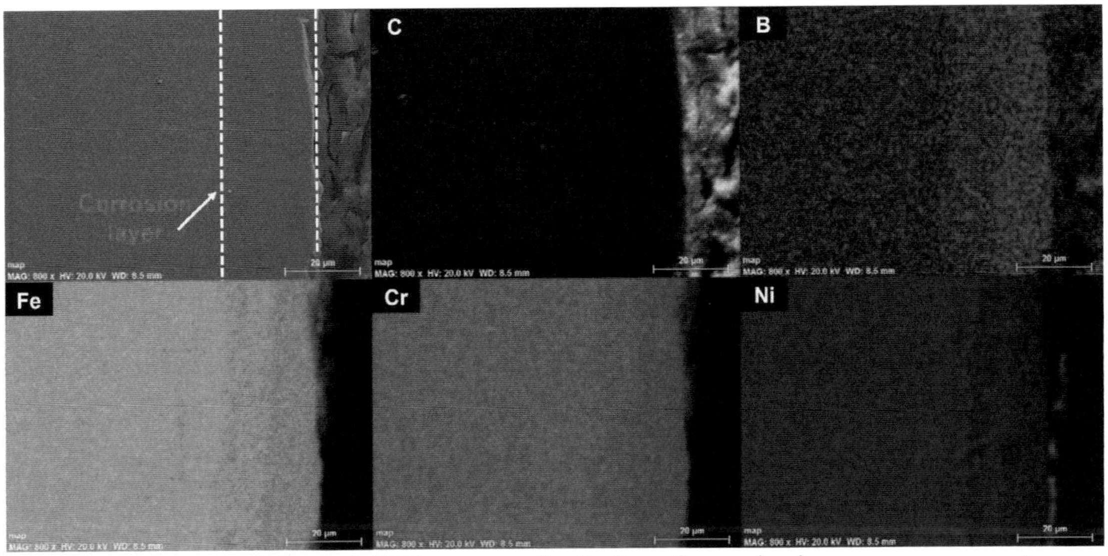

Figure 6. EDS mapping distribution of corrosion layer

For the 15-15Ti/Na/B_4C ternary system, B_4C decomposes and dissolves into sodium at elevated temperatures. Then, the Fe and Cr elements of 15-15Ti react with B dissolved in sodium resulting in the formation of boride of Fe and Cr. According to the thermodynamic data, the heat formation of FeB and Fe_2B is greater than that of CrB, and heat formation of CrB is greater than that of Ni_3B. FeB, Fe_2B and CrB is more stable than boride of Ni. Therefore, due to the reaction among Fe, Cr, Ni, B and C at the interface between corrosion layer and 15-15Ti matrix with the formation of boride of Fe and Cr, the Ni element is compelled from corrosion layer to the inner surface of 15-15Ti cladding tube.

4. Conclusion

For the 15-15Ti/Na/B4C ternary system, the compatibility experiment was performed to study the alloy elements migration behaviour and corrosion procedure with the reaction of the dissolved B and C in sodium. Through the surface and cross section morphology, elements distribution and XRD phase analysis, the conclusions can be achieved as follows.

(1) For the 15-15Ti/Na/B4C ternary system, the B4C decomposes and dissolves into sodium at elevated temperatures.
(2) The boronation of inner surface of 15-15Ti takes place via the reaction with B dissolved into sodium, resulting in the formation of boride of Fe and Cr.
(3) The Ni in 15-15Ti compelled to the surface and precipitates on the surface of cladding tube, and the precipitates grow up with increase of the experiment time.

5. References

[1] Abram T, Ion S. 2008 Generation-IV nuclear power: A review of the state of the science [J]. 2008 *Energy Policy.* **36** 4323-4330.

[2] Prasad Reddy G V, Karthik V, Latha S, Venkiteswaran C N, Ramachandran D, Albert S K. 2021 Core Materials for Sodium-cooled Fast Reactors: Past and Future Prospect [J]. *Materials Performance and Characterization,* **11**(2) 1-51.

[3] Chandran K, Clinsha P C, Lavanya M, Anthonysamy S, Gnanasekar K I. 2018 Chemical Interaction of D9 Alloy Clad with B_4C in Liquid Sodium: Studies Employing XPS, XRD, and SEM [J]. *Journal of the American Ceramic Society*, **102**(5) 2932-2947.

[4] Romedenne M, Rouillard F, Hamon D, Tabarant M, Monceau D. 2019 Chemical Interaction of Austenitic and Ferritic Steels with B_4C Power in liquid Sodium at 600°C [J]. Corrosion: the Journal of Science & Engineering, **75**(10) 1173-1182.

[5] Chandran K, Anthonysamy S, Lavanya M, Sudha R, Reshmi P R, Annie D, Madhavan R R, Prasanthi, T N, Sudha C, Saroja S and Ganesan V. 2014 Studies on the Chemical Compatibility of Alloy D9 with B4C in the Presence of Sodium [J]. *Procedia Engineering* **86** 631- 638.

[6] Heuvel H J, Holler P, and Donner P. 1985 Absorber Material Cladding Chemical Interaction in Vented Fast Breader Reactor Absorber Pins. B4C/Stainless Steel Chemical Interaction in Sodium Environment and Effect of Metallic Nb and Cr Layers [J]. *Journal of Nuclear Materials,* **130** 517-523.

Journal of Physics: Conference Series **2749** (2024) 012011 doi:10.1088/1742-6596/2749/1/012011

Direct formation of nano-scale dislocation loops during displacement cascade in BCC Fe studied by molecular dynamics simulation

Han Cao[1,*], Xinfu He[1,*], Shi Wu[1], Dongjie Wang[1], Lixia Jia[1], Yankun Dou[1], Wen Yang[1]

[1] China Institute of Atomic Energy, Beijing 102413, China

[*]Corresponding authors.

E-mail addresses: zcaohan@foxmail.com (Han Cao), xinfuhe@gmail.com (Xinfu He)

Abstract. In this work, the displacement cascades in bcc iron were simulated by molecular dynamics (MD). The formation of dislocation loops was found in cascade simulations when the energy of PKA (EPKA) was above 20 keV. For the first time, we reported the formation of <100> interstitial dislocation loops in cascade simulation of bcc iron under low EPKA, about 40 keV, much lower than that in Ref. 10. On the other hand, our results suggest that <100> loops all transformed into 1/2<111> loops within 200 ps. This will help us build a microscopic image and mechanism about the formation of interstitial dislocation loop.

1. Introduction

Alpha-Fe-based material is an important structural material in nuclear reactors. Embrittlement at low temperature is a major problem for this kind of materials under neutron irradiation. The key reason for irradiation embrittlement is matrix damage [1] originated from the collision between the incident particles and lattice atoms [2]. The collision leads to cascade, which will produce considerable defects, including vacancy clusters, interstitial clusters, and dislocation loops. Among them, dislocation loops will contribute greatly to the irradiation hardening and embrittlement of materials.

As for the interstitial dislocation loops produced in cascades, <100> interstitial dislocation loops were observed in irradiated bcc-Fe based materials while 1/2<111> interstitial dislocation loops have only been found after irradiation in other bcc metals. In comparison, 1/2<111> interstitial dislocation loops are mobile, so they tend to annihilate at the interfaces. Meanwhile, <100> interstitial dislocation loops are immobile at T < 400 ℃ [3], therefore their contribution to the hardening is much more than 1/2<111> dislocation loops. However, the formation mechanism of <100> interstitial dislocation loop has been an unresolved problem since the 1960s, and there are various possible formation mechanisms. First, Eyre et al. proposed the stacking fault mechanism that <100> loop can be formed through the <110> stacking faults' reaction[4]. However, because of the high stacking fault energy in BCC metals, this mechanism is considered to be wrong. The second one is dislocation loops' reaction mechanism. Marian[5] and Arakawa[6] respectively studied two 1/2 <111> dislocation loops' reaction by MD method and transmission electron microscopy experiments, and there was no evidence of forming a complete and stable <100> loop yet. In 2013, Xu[7] used kinetic monte carlo (KMC) method to simulate the dislocation loops' reaction, and their results showed that the probability of forming <100>

loops was only 27.7%, and it was based on the assumption that two initial dislocation loops could be encountered. Thus, dislocation loops' reaction was obviously not the main formation mechanism of <100> loops. The third one was the transformation mechanism of dislocation loops, which was first found by Arakawa et al.[8] in the electron irradiation experiment. However, the transformed < 100> loop were very unstable, and it turned back to 1/2 <111> loop in a short period of time.

As we all know, cascade is an important path to form the interstitial dislocation loops in bcc iron. Cascades occur at atomic scale and finish in about 10 picoseconds (ps), which is not suitable for experimental research. Therefore, we can consider computational simulation. In the multi-scale simulation of materials, there are many methods, such as first principles, MD, KMC, rate theory (RT), finite element simulation and so on. According to the time and spatial scale of cascades, MD is chosen as the main method to study cascades. At the same time, cascade simulation has important guiding significance for the understanding of irradiation mechanism. Plenty of cascade simulations can provide the necessary parameters for other microstructural and mesoscopic simulation methods (KMC, RT), such as the defect production efficiency and the clustered fraction.

Recently, the simulation study of cascades has attracted the attention of researchers. In 2017, Granberg [9] carried out several cascade simulations in pure iron and Fe-Cr alloys and found the generation of the <100> interstitial dislocation loop after a thousand iterative simulations. Recently, Peng' study [10] shows that MD simulations of cascades can produce <100> interstitial dislocation loops when EPKA is up to 200 keV. In this study, we will make some attempts to find <100> interstitial dislocation loops in single cascade under low EPKA.

2. Simulation methods

In this work, the interatomic potentials appropriate for crystalline and liquid iron [11], which was established by Mendelev in 2003, is used to simulate the cascade in bcc iron materials by the MD code Large-scale Atomic/Molecular Massively Parallel Simulator (LAMMPS) [12]. Meanwhile, M07 potential [13,14] is used to verify the simulation results about dislocation loops. There are two sizes of simulation boxes according to EPKA. The specific corresponding box size for each EPKA is shown in Table 1. The simulated irradiation temperature is set at 600 K, and the three dimensional periodic boundary conditions are used in every simulation [15]. The system is heated to the irradiation temperature before the cascade and insulated for 20ps to ensure the system to achieve thermal stability.

Table 1 Number of atoms in simulation box for different E_{PKA}

E_{PKA} (keV)	1~30	40、50
Atoms in box	1024000 (80×80×80×2)	6750000 (150×150×150×2)

In cascade, PKA must be introduced into the simulated box. One PKA is put in center of every box through three directions, <122>, <135> or <235> [15]. EPKA is represented by different initial velocities. In order to achieve the statistical effect of the simulation results, each simulation condition is calculated nine times. The total simulation time varies according to EPKA to ensure enough annealing of cascades, which is shown in Table 2. In cascades simulation, the simulation time step and the frequency of the output atom position information are changed at different stages, as follows: At the beginning of cascades, the time step is 0.0001 ps, and the position information of all atoms is output every 0.05 ps. After 5.5 ps, the time step is changed into 0.001 ps, and the information is output once every 0.5 ps.

Table 2 Simulation time for different E_{PKA}

E_{PKA} (keV)	1~20	25、30	40、50
Time (ps)	15.5	25.5	35.5

After the simulation of the cascade finished, the Wigner-Seitz cell method [16] is used to identify the SIA (self-interstitial atom) and vacancies, and the type of defects is determined according to the number of atoms in the cell. The number of defects in each frame of each simulation is counted. In this work, cluster analysis of the cascade simulation results above 20 keV energy PKA is carried out. The critical radius of the SIA cluster is second near neighbor distance and the critical radius of the vacancy cluster is third near neighbor distance [17]. Dislocation analysis (DXA) method [18] in The Open Visualization Tool (OVITO)[19] is used to analyze dislocation loops after simulation.

3. Results and discussion

3.1. cascade simulation

Fig. 1 depict snapshots of the evolution of defects generated during cascade caused by 50 keV PKA in bcc iron at 600 K. In this cascade, the evolution of defects went through four stages: defect generation, the emergence of the peak, the annihilation of defects, and the final formation of stable defects. It can be seen that after the formation of a 30 keV PKA, a large number of point defects are generated near the PKA within 0.1 ps; At around 1.0 ps, the number of Frenkel defect pairs reaches its peak; Afterwards, after about 14.0 ps of defect recombination, three 1/2<111> interstitial-type dislocation loops, one <100> vacancy-type dislocation loop, clusters of different sizes, and multiple point defects distributed in the matrix were ultimately generated within the system.

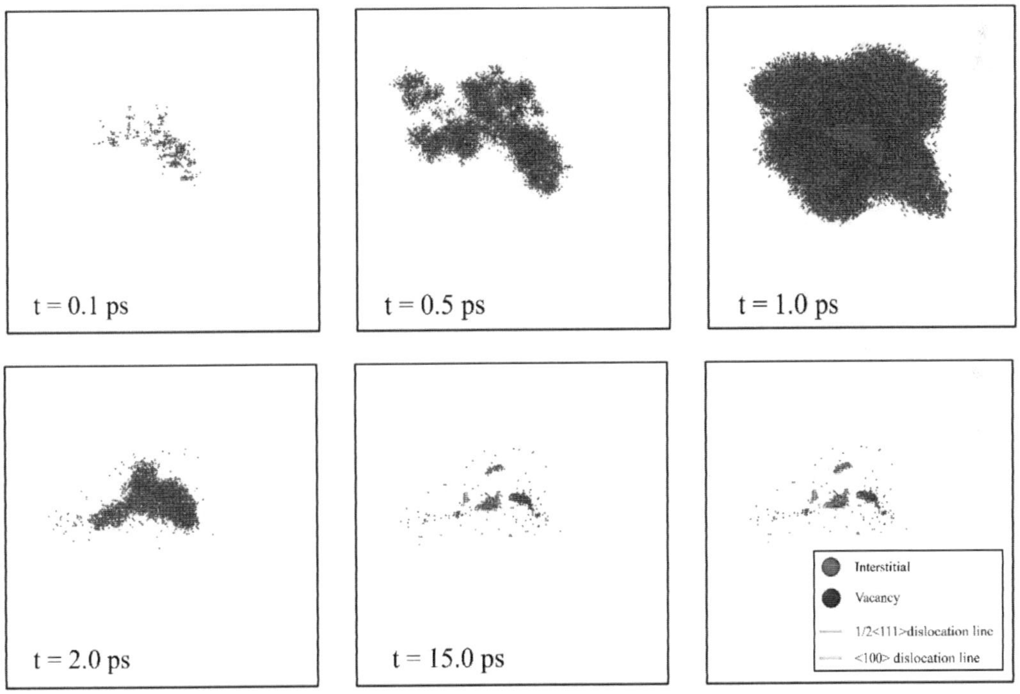

Fig. 1. Visual screenshot of defect evolution during cascade of 50 keV PKA in BCC iron at 600 K

Fig. 2 quantitatively gives us the evolution information of the number of defects in cascades caused by different EPKA at 600 K, the peak number of defects and time value of reach peak become lager with the increase of EPKA, as shown in Fig. 3.

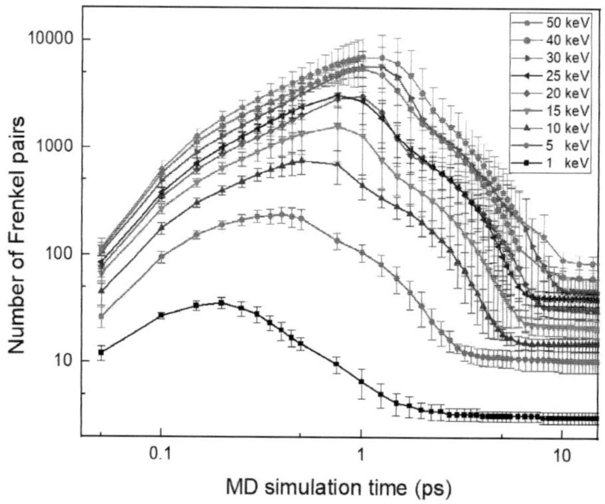

Fig. 2. The number of defects in 600 K cascades with time evolution

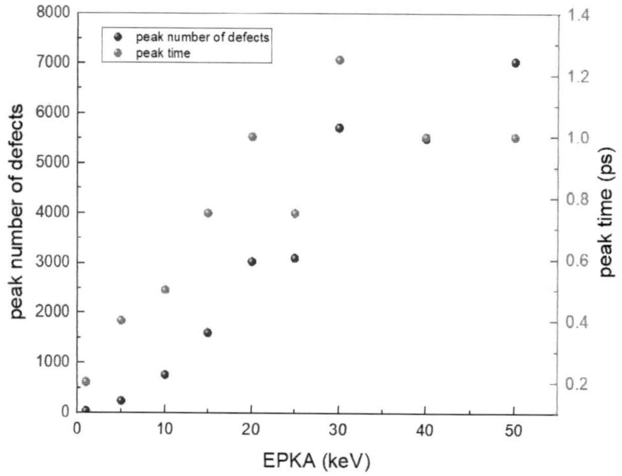

Fig. 3. The peak number of defects and time value to reach peak with EPKA

Fig. 4 shows average number of surviving defects in cascades caused by different EPKA at 600 K, and these data are quite different from the predicted value of the Norgett-Robinson-Torrens (NRT) model [20]. This phenomenon has been discovered by many researchers for decades, and the reason for overestimation is that the NRT model does not consider the recombination of the displacement atoms. Recently, Nordlund et al. [21] proposed the athermal recombination corrected dpa (arc-dpa) model, which can fix the overestimation of NRT model. It can be seen from Fig. 3 that the results obtained in this study are consistent with the predicted values of arc-dpa model and are similar to the results of other simulation studies of cascade in bcc iron. It is noted that the number of surviving defects is large under high EPKA (>20keV) conditions, and it is possible that large clusters have been formed. Therefore, cluster analysis and dislocation analysis are performed on the results, that will be discussed in detail in the following two subsections.

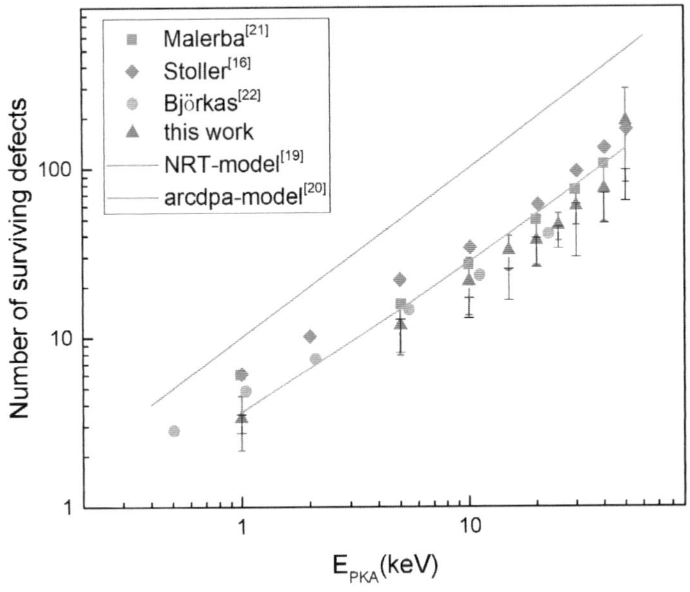

Fig. 4. Average number of surviving defects in cascades at 600 K

3.2. Cluster analysis

Fig. 5 and Fig. 6 are the three-dimensional columnar graphs of the SIAs and vacancies clusters' size and number after cascades caused by different EPKA at 600 K. The number of clusters produced in cascades caused by energy less than 20 keV PKA is small, which is not significant statistically, so they are not shown in the graph. It can be seen from the graph that the higher the EPKA is, the larger the probability of large clusters and the larger the total number of clusters will be. Fig. 7 and Fig. 8 show the ratio of interstitials and vacancies in clusters to total surviving defects at 600 K. It is shown that with the change of the EPKA, the clustered fraction of interstitials and vacancies did not change too much at 600 K, and it was stable at around 40 % and 50 % respectively. It is noted that when the EPKA is high, there are clusters containing tens or even hundreds of interstitial atoms and vacancies. Dislocations are suspected to be formed, so dislocation analysis is carried out for these special cases in next section.

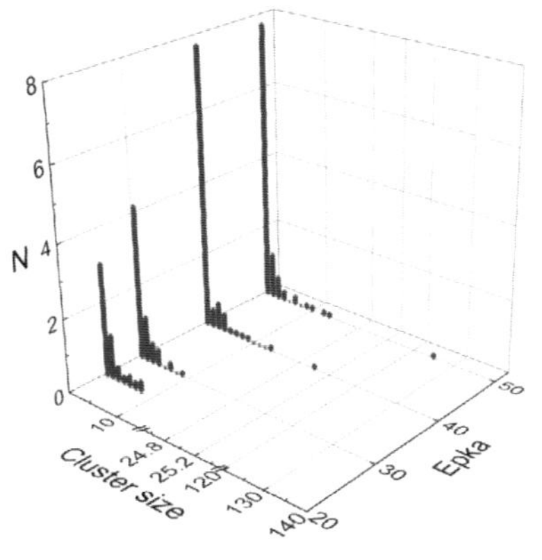

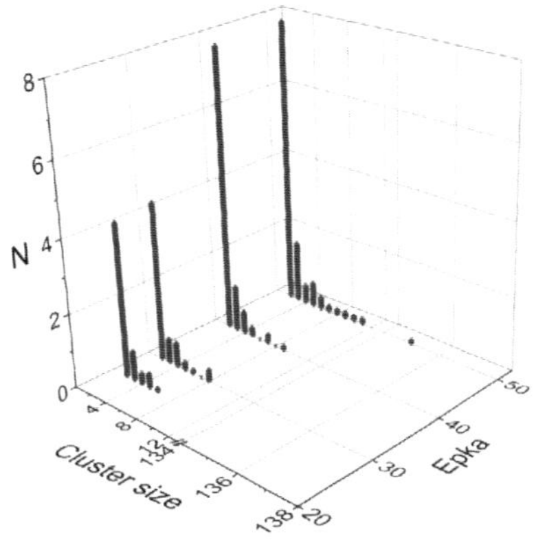

Fig. 5. Size and number of interstitial clusters Fig. 6. Size and number of vacancy clusters

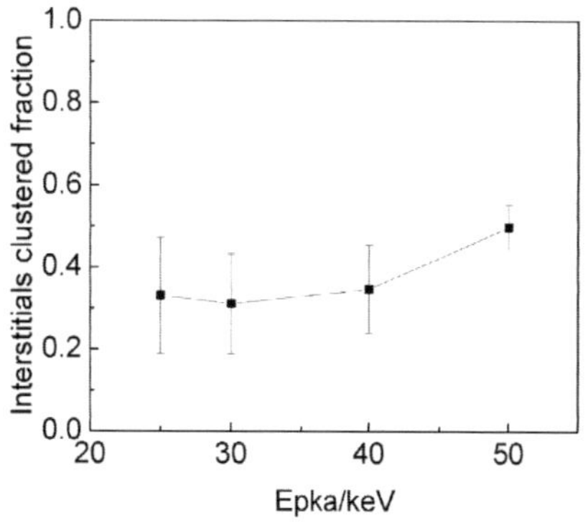

Fig. 7. Interstitials clustered fraction

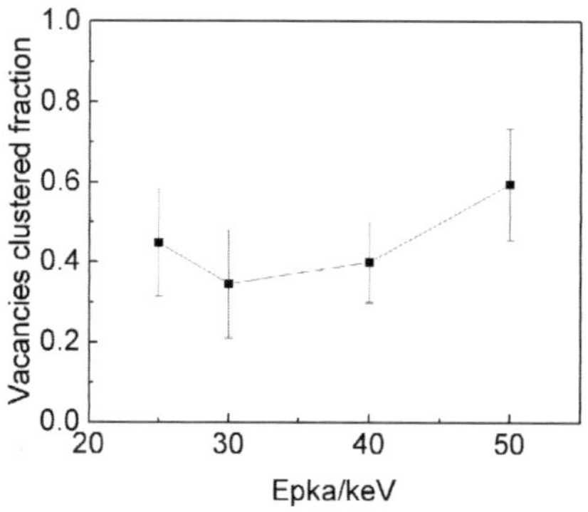

Fig. 8. Vacancies clustered fraction

3.3. Dislocation loop analysis

In the cascade simulation caused by some cases, the dislocation loops are found, and these dislocation loops correspond to the large size clusters in the column Fig. 5 and Fig. 6.

Cascades produced dislocation loops composed of a large number of interstitial atoms or vacancies. By analyzing the results of 40 and 50 keV PKA cascades, we found dislocation loops with diameter from 0.8 to 10 nm in 9 examples, two types of dislocation loops are mainly formed, namely 1/2<111> interstitial-type dislocation loops, and <100>vacancy-type dislocation loops. In interstitial-type dislocation loops, as shown in Fig. 2, the relative directions of the two nearest neighboring atoms are mostly<111>, with a small amount being<112>and other directions. This is because when a large number of self-interstitial atoms gather, the cluster composed of dumbbells in the<111>direction has a higher binding energy [22]; In vacancy type dislocation loops, the relative directions of the two nearest vacancies are mostly in the<100>direction.

However, one of these loops was an interstitial's dislocation loop with Burgers vector of <100> when EPKA was 40keV. After the time delay of the simulation, it was found that the <100> interstitial-type dislocation loop transformed into 1/2<111> loop completely as shown in Fig.9. This dislocation loop appeared at 1.25 ps and maintained at <100> Burgers vector around 200 ps. In addition, the direction of dumbbell in dislocation loops also changed from <100> to <111> during transformation. To verify the universality of the results, we also simulated cascades with M07 potential, and three of them produced <100> interstitial-type loops. These three <100> loops all rapidly transformed into 1/2 <111> loops within 100 ps. In the four simulations appearing <100> dislocation loops, the minimum EPKA was 20 keV and the maximum was 50 keV, which was significantly lower than the 200 keV required by Peng's research [10].

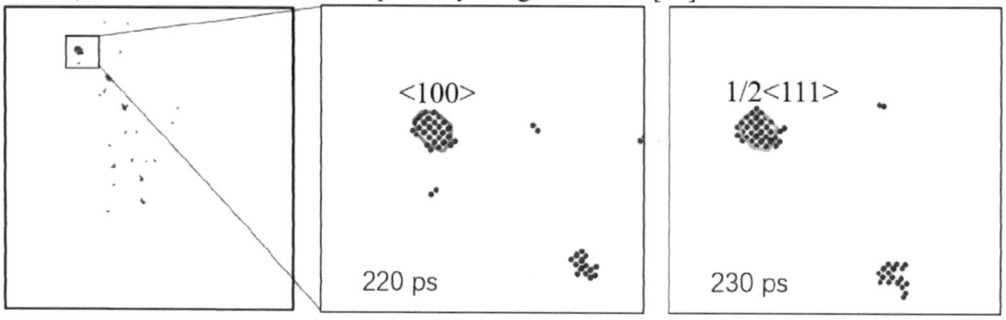

Fig. 9. Change of Burgers vector of dislocation loop

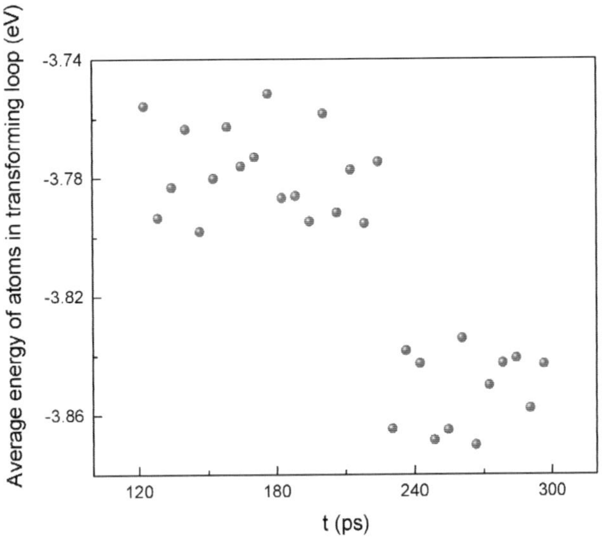

Fig. 10. Change of energy of dislocation loop

According to the classical dislocation theory, we know that the energy of the dislocation is proportional to the square of the Burgers vector. Through simple calculation, we know that for the same size of 1/2<111> and <100> interstitial's dislocation loop, the former energy is 3/4 of the latter. To this end, we use the results to make a qualitative verification of the theoretical results: The nature of dislocation loops generated by cascades are large clusters, which contains many interstitial atoms or vacancies. Fig. 10 shows the change of average energy of interstitial atoms in this dislocation loop before and after the transformation. The energy of the dislocation loop from <100> to 1/2<111> is reduced, which is the reason why <100> dislocation loops can transform into 1/2<111> loops.

4. Conclutions

In this work, based on MD, a series of cascades simulation in bcc iron was studied, and the following conclusions can be drawn:

Dislocation loops with different sizes and types were found in cascades caused by PKA with energy varying from 20 keV to 50 keV, containing several <100> dislocation loops. These <100> loops all transformed into 1/2<111> loops within 200 ps, because the energy of <100> dislocation loops is higher than 1/2<111> dislocation loops with the same size, which lays the foundation for the study of the formation mechanism of <100> interstitial's dislocation loops in bcc iron.

5. References

[1] Lambrecht M, Malerba L, Almazouzi A. Influence of different chemical elements on irradiation-induced hardening embrittlement of RPV steels[J]. Journal of Nuclear Materials, 2008, 378(3): 282-290.

[2] Was G S. Fundamentals of radiation materials science: metals and alloys[M]. Springer, 2007.

[3] Yao Z, Jenkins M L, et al. The temperature dependence of heavy-ion damage in iron: A microstructural transition at elevated temperatures[J]. Philosophical Magazine, 2010, 90(35-36): 4623-4634.

[4] Eyre B L, Bartlett A F. An electron microscope study of neutron irradiation damage in alpha-iron[J]. Philosophical Magazine, 1965, 12(116): 261-272.

[5] Marian J, Wirth B D, Perlado J M. Mechanism of formation and growth of <100> interstitial loops in ferritic materials[J]. Physical review letters, 2002, 88(25): 255507.

[6] Arakawa K, Amino T, Mori H. Direct observation of the coalescence process between nanoscale dislocation loops with different Burgers vectors[J]. Acta Materialia, 2011, 59(1): 141-145.

[7] Xu H, Stoller R E, Osetsky Y N, et al. Solving the puzzle of <100> interstitial loop formation in bcc iron[J]. Physical review letters, 2013, 110(26): 265503.
[8] Arakawa K, Hatanaka M, Kuramoto E, et al. Changes in the Burgers vector of perfect dislocation loops without contact with the external dislocations[J]. Physical review letters, 2006, 96(12): 125506.
[9] Granberg F, Byggmästar J, Sand A E, et al. Cascade debris overlap mechanism of <100> dislocation loop formation in Fe and FeCr[J]. EPL, 2017, 119(5): 56003.
[10] Peng Q, Meng F, Yang Y, et al. Shockwave generates <100> dislocation loops in bcc iron[J]. Nature communications, 2018, 9(1): 4880.
[11] Mendelev M I, Han S, Srolovitz D J, et al. Development of new interatomic potentials appropriate for crystalline and liquid iron[J]. Philosophical magazine, 2003, 83(35): 3977-3994.
[12] Plimpton S. Fast parallel algorithms for short-range molecular dynamics[J]. Journal of computational physics, 1995, 117(1): 1-19.
[13] Malerba L, Marinica M C, Anento N, et al. Comparison of empirical interatomic potentials for iron applied to radiation damage studies[J]. Journal of Nuclear Materials, 2010, 406(1):19-38.
[14] Marinica M C, Willaime F, Crocombette J P. Irradiation-induced formation of nanocrystallites with C15 laves phase structure in bcc iron[J]. Physical review letters, 2012, 108(2): 025501.
[15] He X. Molecular dynamics simulation of matrix radiation damage in Fe-Cu alloy[J]. Acta Metall Sin, 2011, 47(7): 954-957.
[16] Kittel C, McEuen P, McEuen P. Introduction to solid state physics[M]. New York: Wiley, 1976.
[17] Stoller R E, Odette G R, Wirth B D. Primary damage formation in bcc iron[J]. Journal of Nuclear Materials, 1997, 251(6):49-60.
[18] Stukowski A, Bulatov V V, Arsenlis A. Automated identification and indexing of dislocations in crystal interfaces[J]. Modelling and Simulation in Materials Science and Engineering, 2012, 20(8): 085007.
[19] Stukowski A. Visualization and analysis of atomistic simulation data with OVITO–the Open Visualization Tool[J]. Modelling and Simulation in Materials Science and Engineering, 2009, 18(1): 015012.
[20] Norgett M J, Robinson M T, Torrens I M. A proposed method of calculating displacement dose rates[J]. Nuclear engineering and design, 1975, 33(1): 50-54.
[21] Nordlund K, Zinkle S J, Sand A E, et al. Improving atomic displacement and replacement calculations with physically realistic damage models[J]. Nature communications, 2018, 9(1): 1084.
[22] Terentyev D, Hou M. Study of radiation effects in FeCr alloys for fusion applications using computer simulations[J]. 2006.

Acknowledgments
This research was funded by National Key Research and Development Program of China under the project number 2019YFB1900902 and 2023YFB3002100.

Fabrication and properties of SiC-based accident-tolerant fuel with precisely-regulated TRISO particles dispersion

Zongyi Shao[1,2], Yumeng Zhao[1,2 *], Shuaishuai Feng[1,2], Ying Meng[2], Chen Chen[2], Wentao Lui[1,2], Wei Liu[1,2] and Zhenfang Cai[2]

[1] CNNC Key Laboratory of Fabrication Technology of Reactor Irradiation Special Elements, Baotou 014035, China
[2] China North Nuclear Fuel Element Co., Ltd., Baotou 014035, China
Email: zhaoym14@tsinghua.org.cn (Y. Zhao)

Abstract The ceramic-based coated particle dispersed fuel (CDF) has arrested worldwide attention as a kind of promising accident-tolerant fuel since the Fukushima-Daiichi accident in 2011. To improve safety and in-pile performance, regularly-distributed CDF was formulated. A novel fabrication method was developed to accurately target TRistructural-ISOtropic (TRISO) particles to preset positions in the SiC matrix via a self-developed precise TRISO-controlled moulding equipment. After sintering, the samples were observed by SEM and X-μCT, which showed that the TRISO particles were uniformly and regularly arranged in the SiC matrix, and were almost intact with little extrusion and breakage. The density, thermal conductivity, and thermal shock performance of the prepared samples were examined and proved to be consistent with those of the pellets with randomly-distributed TRISO particles, which could satisfy the safety requirements. Small batch manufacturing was realized and production efficiency was enhanced. This new technology offers a promising pathway to produce high-reliability ceramic-based CDF with precisely-located TRISO particles, which promote the development of advanced accident-tolerant fuels.

1. Introduction

After the Fukushima-Daiichi accident in 2011, increasing international efforts have been paid to the accident-tolerant fuel (ATF) with accident tolerance and inherent safety [1–3]. The ceramic-based coated particle dispersed fuel (CDF), with TRistructural-ISOtropic (TRISO) fuel particles dispersed in ceramic matrix, has become a candidate ATF for advanced light water reactors (LWRs) due to its excellent fission products containing capability, outstanding corrosion resistance of the matrix and high thermal conductivity [4–7].

The TRISO particles were first developed in German [8], and the concept of TRISO particles dispersed fuel with SiC matrix was put forth by Oak Ridge National Laboratory (ORNL, America) [9]. The TRISO particle with a diameter of nearly 1.0 mm is typically composed of a fissionable microsphere (e.g., UO_2) kernel, surrounded successively by layers of porous pyrolytic carbon buffer, dense inner pyrolytic carbon (IPyC), silicon carbide (SiC), and dense outer pyrolytic carbon (OPyC) [10,11]. The kernels are prepared by the sol-gel method, and the cladding layers are prepared by fluidized bed chemical vapor deposition (FB-CVD) [12]. TRISO particles were supposed to have the advantages of high thermal conductivity and good irradiation stability, which could effectively prevent the release of fission products.

SiC matrix with high thermal conductivity and outstanding in-service performance can effectively reduce the centre temperature of the pellets, and protect the TRISO particles in the matrix from the loading and corrosion damage. However, based on the nature of strong covalent bonding in SiC materials, there is only a trace amount of neck growth between SiC powders (3 ~ 5 μm) without volume shrinkage densification during pressureless solid-phase sintering, even at very high temperatures. Therefore, it is necessary to introduce higher temperatures, higher mechanical pressure, or additives to complete the sintering of SiC ceramics.

In the traditional CDF production process, TRISO particles are first overcoated to avoid direct contact with each other, and blended with SiC matrix to acquire homogeneous distribution, and then moulded and sintered to the desired performance. The positions of TRISO particles are randomly distributed in matrix and by no means be specific controlled. Nowadays, a few studies have achieved regularly-arranged TRISO particles in the matrix by employing state-of-the-art technologies [13–17]. Chao L et al. [13,14] introduced stereolithography-based addictive manufacturing (SLAM) to produce the $MgO\text{-}PuO_2$ dispersion nuclear fuel pellets to ensure a highly uniform distribution of the nuclear fuel microspheres, where the MgO matrix acted as a structural material that holds pellets in their three-dimensional shape. Ang C et al. [15,16] proposed a moulding method for stacking preforms equipped with radially reinforced TRISO particle planes into cylindrical pellets. Liu C et al. [17] fabricated multilayer fully ceramic microencapsulated fuel pellets using tape casting, with TRISO particles in an orderly arrangement. These attempts, however, were complex and time-consuming, and difficult to be applied in industrial production by now.

In this work, an innovative fabrication method is present to produce SiC-based CDF pellets, where the TRISO particles were accurately targeted to preset positions in the SiC matrix via a self-developed precise TRISO-controlled moulding equipment. After spark plasma sintering (SPS), the microstructure of the prepared samples was observed by scanning electron microscope (SEM) and X-ray micro-computed tomography (X-ray μCT). The density, thermal conductivity, and thermal shock performance of the samples were examined and compared with randomly-distributed TRISO particles. The pellets produced by this method showed satisfying microstructure and performance, which provided a prospective technical solution for accident-tolerant fuels.

2. Methods

2.1. Materials

The matrix material used in this study was high-purity nanoscale 3C-SiC (β-SiC) powder with granularity of 70 ~ 90 nm. According to previous studies, a certain quantity of sintering aids could promote the densification sintering of SiC [18]. Thus, the SiC powder was mixed with 8 wt.% sintering aids (Al_2O_3 and Y_2O_3, with weight ratio of 1:1 and granularity of 60 ~ 70 nm) by wet ball milling in the agate mill jar for 4 ~ 48 h. The milled powder was then dried, broken and sieved to achieve the mixing powder for SiC matrix. The coated particles with Al_2O_3 kernel and four SiC and/or pyrolytic carbon layers were used to simulate nuclear TRISO particles.

2.2. Sample preparation

The samples were prepared by compacting SiC matrix powder and single-layer regularly-arranged TRISO particles layer by layer, to form the global precisely-regulated CDF pellets. This process was carried out by a self-developed precise TRISO-controlled moulding device, which was composed of conveyor belt, manipulator and motor to manipulate and transport the moulds and materials.

The detailed process of pellets preparation is shown in Figure. 1. A layer of mixed SiC powder was first put in the bottom of the shaping mould. The regular arrangement of TRISO particles was accomplished by a particle-sucking unit, where the TRISO particles were taken out by negative pressure. This unit moved to a certain height above the mould, and withdrew the negative pressure and the TRISO particles fell on the SiC powder. After that, one more layer of SiC powder was spread on the particles. A fan was employed to smooth the surface of each SiC powder layer to avoid the accumulation of SiC powder. The mould was then transported to the press station for compacting. The cycles of dropping TRISO particles, putting SiC powder and compacting were repeated until the total height of SiC and TRISO layers reached the requirement. The weight and height of SiC powder for each layer was calculated according to the dimension and TRISO particles contents of the designed pellets.

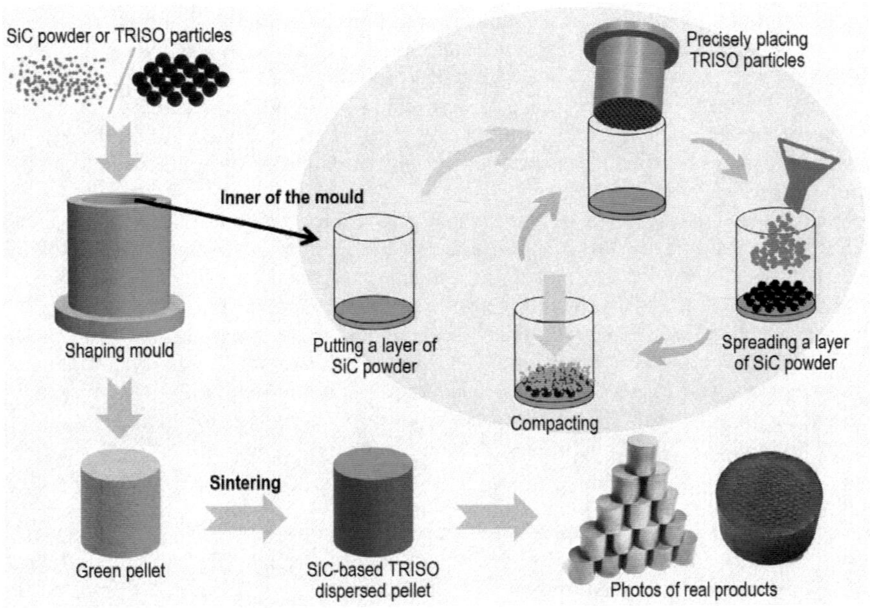

Figure 1. The preparation process of the precisely-regulated CDF pellets.

The regular distribution of TRISO particles in a single layer was affected by multiple factors, including the stirring speed of the fan, the magnitude of vacuum suction, the particles-falling height, the compression force of SiC powder and TRISO particles, etc. These factors have been carefully adjusted to realize the regulated arrangement of TRISO particles. For example, the vacuum suction was set to be -35 ~ -25 kPa to ensure full adsorption of TRISO particles without destroying the pellets, as shown in Figure. 2 (a). The particles-falling height was 0 mm when the TRISO particles were evenly and neatly arranged. as shown in Figure. 2 (b). Overall, the TRISO particles were regularly located in single layer although a few vacant positions existed.

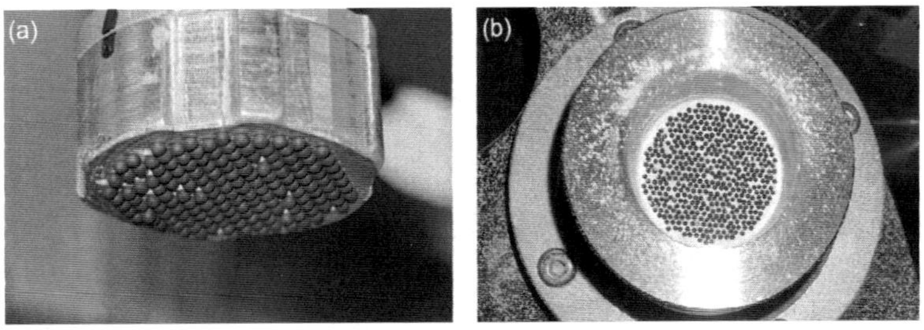

Figure 2. (a) The TRISO particles absorbed when the suction was -35 kPa. (b) The TRISO particles layer when the particles-falling height was 0 mm.

The green pellets were sintered by HPD25-SI spark plasma sintering furnace (FCT Systeme GmbH, German) using graphite dies. The samples were first heated to 1400 °C and kept 3 minutes, and then heated to 1800 °C and kept 10 minutes. The sintering pressure was 30 MPa. The heating and cooling rate were both 100 °C/min.

2.3. Characterization

The surface and internal morphology of the samples was observed by Merlin Compact scanning electron microscope (SEM, Carl Zeiss, German). The interior microstructure, especially the arrangement of TRISO particles, was further explored by Xradia620 Versa X-ray micro-computed tomography (X-ray μCT, Carl Zeiss, German). The X-ray μCT data were analysed by Avizo software.

The density of the samples was detected by the Ultrapycnometer 1000 density analyser (Quantachrome, American), which measured bulk density of samples based on Archimedes's principle and Boyle's law.

The specific heat was measured by the comprehensive thermal analyser, and thermal diffusion coefficients were measured by the LFA457 laser-pulsed thermal conductivity meter (NETZSCH, German). Thermal conductivity responds to the ability of a substance to conduct heat. According to Fourier's law, it was defined as the amount of heat transferred by a unit temperature gradient through a unit thermal conductive surface in a unit time. The thermal conductivity in this study was determined by the product of the specific heat, the thermal diffusion coefficient, and the density of the sample.

The thermal shock performance tests were conducted to explore structural integrity of fuel pellets under extreme in-pile condition. The samples were heated to 1200 °C, kept for 10 minutes, and shocked with room temperature (R.T) water to observe cosmetic integrity.

3. Results and analysis

The precisely-regulated CDF pellet samples were successfully produced following the process in Figure. 1. The real cylindrical products with diameter of ~ 21.6 mm and height of ~ 22.2 mm are also shown in Figure. 1 (The left ones of the photos of real products). The right one was specially prepared with less powder on the top layer to directly observe the distribution of TRISO particles. The designed width of fuel-free zone was ~ 0.8 mm, and volume content of TRISO particles was ~ 25%.

3.1. Microstructure and distribution of TRISO particles

The regular arrangement of TRISO particles in SiC matrix was the key point in this study. A single layer of TRISO particles dispersed in SiC observed by SEM is shown in Figure. 3, which is the backscattered electron imaging, and the different components could be distinguished. The arrangement of TRISO particles retained orderly after pressurized SPS sintering, and the distance between the particles was consistently uniform, as shown in Figure. 3 (a). Figure. 3 (b) presented the internal structure and morphology of one typical TRISO particle. The Al_2O_3 kernel was located in the center, and the four layers of buffer PyC, IPyC, SiC and OPyC were clearly identified. The whole particle still maintained a good sphericity with no squishing or morphing.

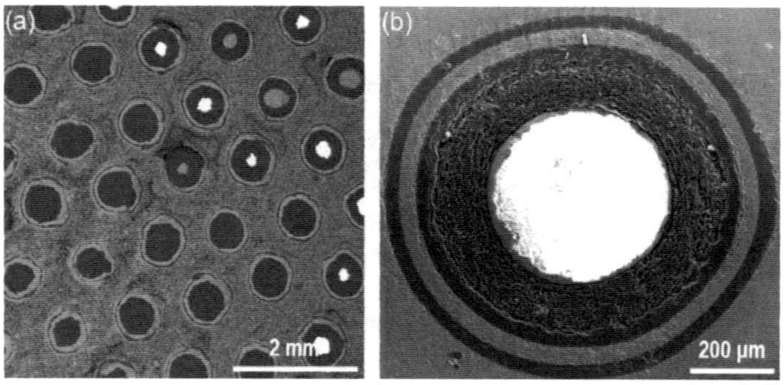

Figure 3. SEM morphology of (a) single-layer TRISO particles distribution, and (b) typical TRISO particle.

The layers of TRISO particle, SiC matrix and their interfaces are shown in Figure. 4. There existed a few pores between Al_2O_3 kernel and buffer layer, as shown in Figure. 4 (a) and (b), which might be due to the looseness of the buffer. The pyrolytic carbon buffer could offer buffering and absorption of fissile debris, accommodate irradiation swelling of the kernel, and hold the fission gas. The interfaces between the buffer and the IPyC, and the interfaces between the IPyC and the SiC, were all relatively closely without conspicuous pores or macro defects. The interfaces between the OPyC and SiC (both SiC layer in the TRISO particle and the SiC matrix, as shown in Figure. 4 (d)), however, showed a porous feature. Researchers have attributed such phenomenon to the mismatched coefficient of thermal expansion of pyrolytic carbon and SiC, or the trapped gas products from reactions at the interface [17,19,20]. Liu C et al. [17] reported a gap along the interface between the OPyC layer of TRISO particles and the SiC matrix after low-pressure (10 MPa) SPS sintering, and recommended higher pressure for sintering to avoid the gap. This might explain why the samples in this study showed only pores but no gaps after sintering by SPS at 30 MPa. With regard to gaseous reaction products, studies have indicated that OPyC could be oxidized during SPS at about 0.1 Pa, leading to gas product of CO/CO_2. Furthermore, the SiC could react with CO/CO_2 and generate SiO (g) and C when the partial pressure of CO/CO_2 was low [20,21]. Therefore, the porous SiC might be caused by cooperative effects of sintering pressure and the trapped gas products during SPS. The region of SiC matrix away from the interface showed consistent pattern with the reported results, including dense ceramic structure and different grain contrast which implied the liquid phase sintering mechanism [17,20,22,23].

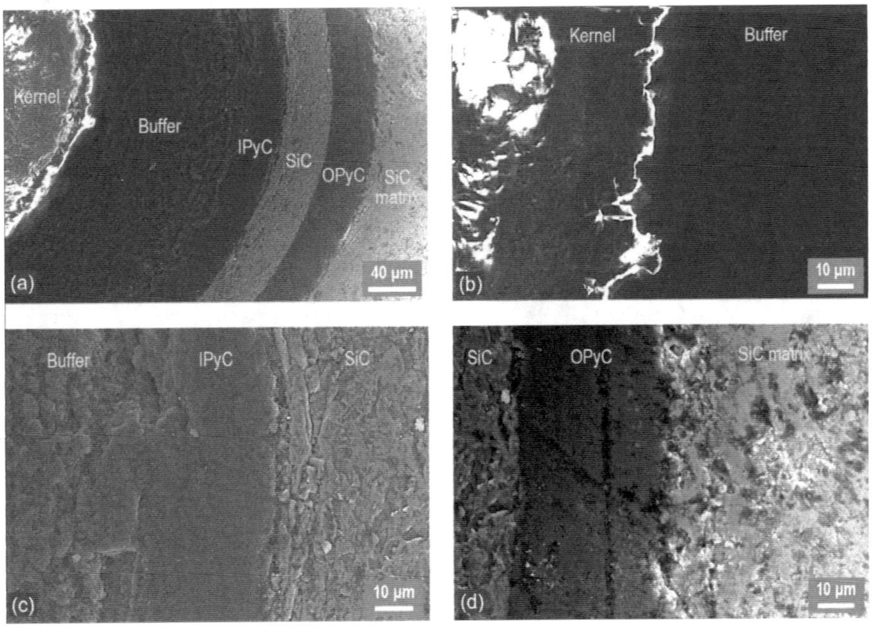

Figure 4. Detailed SEM morphology of typical TRISO particle and SiC matrix. (a) Layers from TRISO kernel to SiC matrix. (b), (c) and (d) are the interfaces between different layers and corresponding microstructure.

The side part of a compacted and demoulded green pellet was broken and the interior particles distribution was exposed, as shown in Figure. 5 (The sample was 22 mm in diameter and 18 mm in height). It could be roughly seen that the boundaries between the TRISO-particle layers were clearly visible, and the layers were evenly spaced, and there was no obvious contact between TRISO particles in neighbouring layers.

Figure 5. Side part of the green pellets.

Figure. 6 shows the X-ray μCT images of precisely-regulated CDF sample. Figure. 6 (a) was taken at a lower magnitude presenting the TRISO particles in a whole pellet, and the uniform distribution of the TRISO particles could be easily identified. A small part of the sintered pellet (Figure. 6 (b)) prepared by wire cutting was also observed, and the limit resolution of the X-ray μCT was about 1/1000 of the sample maximum dimension. The superficial TRISO particles were absent or incomplete due to the damage of wire cutting. It could be induced from Figure. 6 (b) that the TRISO particles were roughly in isometric arrangement within the same layer. This was in accordance with the single-layer distribution of TRISO particles shown in Figure. 6 (d). The spacing between adjacent TRISO particles was measured by Avizo (Figure. 6 (c)) and the mean value was about 280 μm. The three-dimensional distribution of TRISO particles in SiC matrix is demonstrated in Figure. 6 (e). The TRISO particles were intact in appearance, maintaining good sphericity, and did not contact each other.

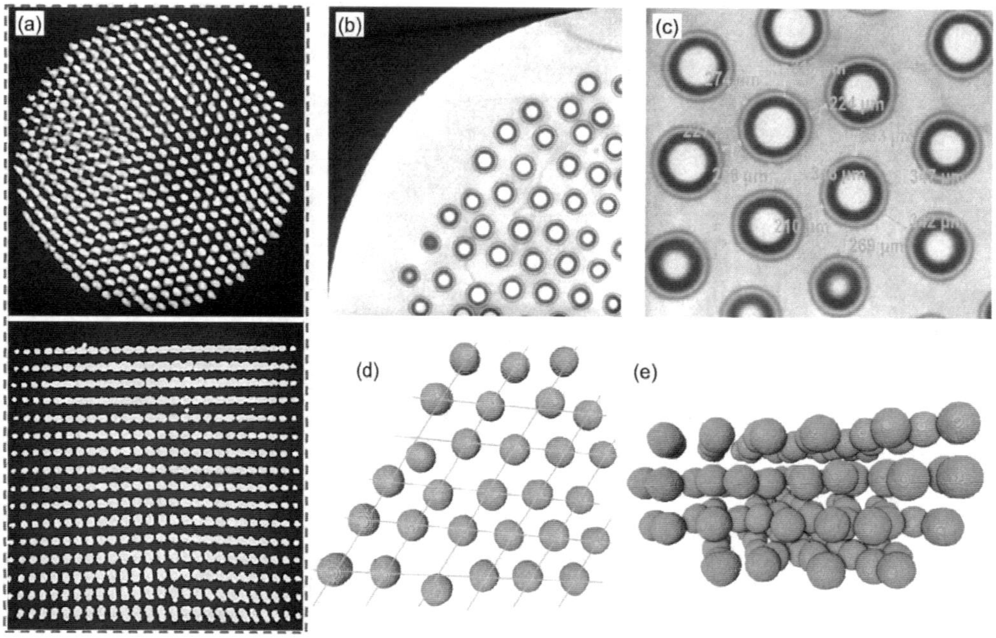

Figure 6. X-ray μCT images of precisely-regulated CDF sample. (a) The X-ray μCT overview of TRISO particles in a whole pellet. (b) Cross-sectional view of the sample. (c) Spacing measurement of regularly-arranged TRISO particles. (d) Single-layer, and (e) multi-layer distribution of TRISO particles in the SiC matrix.

3.2. Thermophysical properties

The measured bulk density of the pellet samples was 2.73 ± 0.03 g/cm^3, approximately 12% lower than that of SiC matrix (3.12 g/cm^3). The thermal diffusion coefficients of multi-layer samples are shown in Figure. 7 (a). The thermal diffusion coefficients decreased with gradually lower rates as temperature rising from R.T. to 1200 °C, and the peak value was about 19.6 ~ 25.6 mm/s^2.

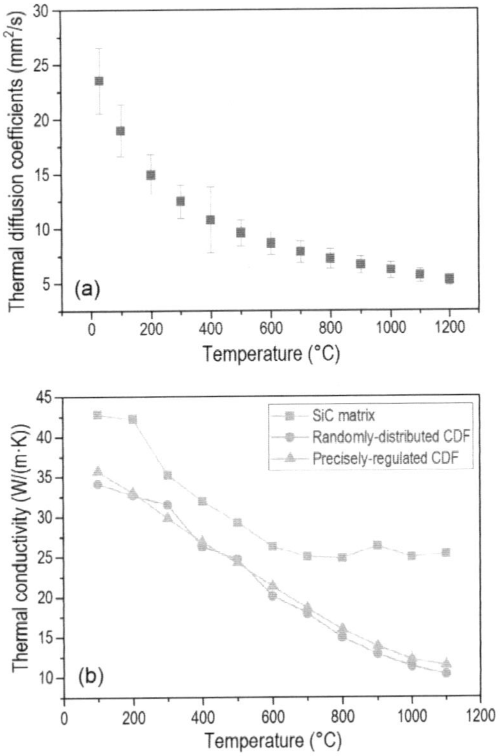

Figure 7. (a) Thermal diffusion and (b) thermal conductivity of CDF samples and SiC matrix.

The thermal conductivity of the precisely-regulated CDF samples is shown in Figure. 7 (b). As the temperature increased from 100 °C to 1100 °C, the thermal conductivity of the samples decreased from 35.7 W/(m·K) to 11.5 W/(m·K), and this trend was consistent with that of the SiC matrix and randomly-distributed CDF samples (also with Al_2O_3 kernel). Compared with SiC matrix, the thermal conductivity of the precisely-regulated CDF samples decreased only by 15% ~ 26% from 100 °C to 700 °C, and decreased by 36% ~ 55% from 800 °C to 1100 °C.

The difference between the thermal conductivity of the precisely-regulated CDF samples and that of the randomly-distributed CDF samples was not significant. On the one hand, this might be due to the fact that the distribution pattern of the TRISO particles in the CDF pellet matrix did not have much influence on their effective thermal conductivity. On the other hand, some TRISO particles were broken and detached from the pellets during the preparation of the test samples by wire cutting, resulting in the testing performance of both samples being close to that of the SiC matrix with a porous surface. Since the volume fraction and thermal conductivity of SiC matrix were larger than that of TRISO particles, it could also be inferred that the distribution pattern of TRISO particles had little effect on the thermal conductivity of the actual pellets, which was mainly determined by the SiC matrix.

Thermal shock tests were carried out on the sintered samples with regular and random TRISO particles distribution, which were heated to 1200 °C and then shocked with water, and the results are shown in Figure. 8 (The samples were 21.6 mm in diameter and 22.2 mm in height). After a thermal shock, both of the samples maintained good integrity, with merely transverse cracks appeared on the surface, and the cracks did not pass through. At the same time, no leakage of TRISO particles was observed, indicating that a thermal shock test would not lead to material leakage, and verifying the good inclusiveness of the CDF pellets as nuclear materials.

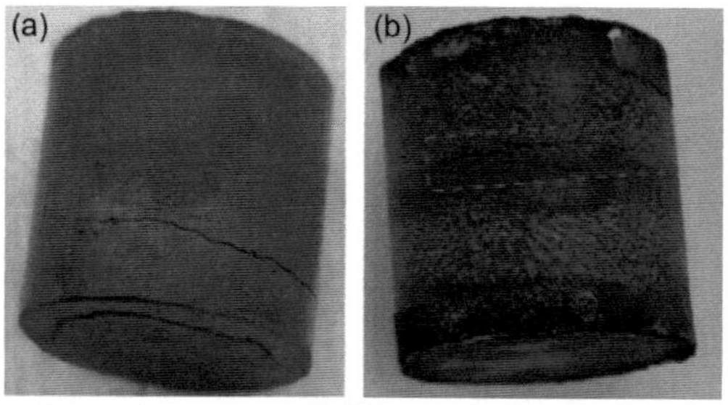

Figure 8. Thermal shock performance of (a) precisely-regulated CDF sample, and (b) randomly-distributed CDF sample

3.3. Batch manufacturing

A turntable module including 6 stations was designed to acquire small batch production of the precisely-regulated CDF pellets. The feeding and unloading devices are shown in Figure. 9. Control experiments preparing samples with 12 layers of SiC powder were conducted. The time consumed for producing one green pellet was about 0.75 hour with productivity of ~1.3 samples per hour. In comparison, it took 2 hours to complete a batch production of 6 green pellets, with productivity of ~3 samples per hour, which was about twice as much as that of single-line production.

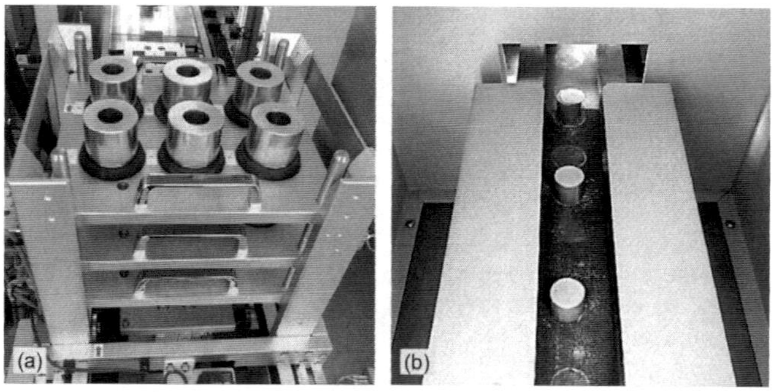

Figure 9. (a) Feeding and (b) unloading devices for small batch manufacturing.

4. Discussion

An effective approach to produce precisely-regulated SiC-based CDF pellets was put forward in this study. The TRISO particles in the prepared samples were proved to be regularly located in general, with little rupture and interaction. The thermal conductivity of these samples was similar to that of randomly-distributed CDF samples, and lower than that of SiC matrix. Researchers have conducted finite element simulations to study the thermal performance of inert (SiC) matrix dispersion pellets [24]. Results showed that the properties of SiC matrix, rather than the shape and distribution of TRISO particles, had more influence on the effective thermal conductivity of the whole pellets. This simulation and other related studies [17] could support the results observed in this study (Figure. 7). Besides, the performance of regularly-arranged CDF pellets might be more similar to the designed fuels, and thus particle distribution pattern could be closer to the ideal state of no contact and breakage.

The present method was convenient to modify the volume contents of TRISO particles. For instance, to fabricate the fuel zone described in this article with diameter of 20 mm, height of 20.6 mm and TRISO particles content of 25 vol.%, 23 layers of TRISO particles and 24 layers of SiC powder were needed to be calculated according to TRISO particle volume and hole numbers of the particle-sucking unit. The SiC powder needed was 0.8 g for each layer calculated in line with SiC density. It could be reasonably referred that the volume contents of TRISO particles could be adjusted by changing the feeding amounts of SiC powder and TRISO particles for each layer. Besides, this fabrication route owned the potential to realize the densest stacking of TRISO particles, which could increase the density of UO_2 in the fuel pellets. Moreover, designing and processing different shaping mould would permit the fabrication of fuel pellets with various shapes (such as cube and hexagonal prism) and dimensions. This straight-forward and step-by-step technology allowed controlling variables to modulate the final products quality.

The current process, however, still could be optimized. Both X-ray μCT and SEM observations revealed further adjustment of sintering method and pressure are needed to better relieve the mismatch of thermal expansion coefficients of different materials. Evaluated sintering pressure could enhance the density of SiC matrix up to 99% T.D. [25], but at the same time more likely to cause deformation and cracks in the case of TRISO particles dispersion.

Comprehensive utilization of conveyor belt, turntable, manipulator and other devices provided possibility for batch manufacturing. The independence and clear delineation between each step of the proposed process allowed the procedure of feeding, particles-falling, demoulding, etc., to be carried out in parallel. Thus, the production efficiency could be increased and this process was expected to be implemented in industrial applications in the future.

5. Conclusion

An application-oriented method to prepare accurately-regulated SiC-based TRISO particles dispersed fuel pellets was established in this study. The concept of processing was intuitive but effective to target the TRISO particles at preset positions. Several highlights of this manufacturing process include: (1) The in-plane and interlayer spacing was uniform according to SEM and X-ray μCT observations, proving successful control of the TRISO particles distribution. (2) Thermal conductivity and thermal shock performance of the produced samples were in consistence with that of traditional CDF pellets, indicating directly applicable capacity of the accurately-regulated CDF pellets. (3) Small batch production was actualized, which could improve efficiency, and lay the foundation for large-scale industrial manufacturing. Regular distribution of TRISO particles could avoid aggregation of particles and alleviate localized heat and stress concentrations, thus improving safety of fuel elements and compatibility with design objectives. The proposed path did not require any complicated techniques, and could be easily adapted to other CDF design schemes. With this new technology, ceramic-based coated particle dispersed fuel with high reliability and precisely-located TRISO particles could be automatedly batch fabricated at low cost, which paved the way for the industrialization of advanced accident-tolerant fuels.

Acknowledgments

The authors wish to acknowledge assistance and encouragement from colleagues. The authors also appreciate the open fund from Center for Testing and Analyzing of Materials (School of Materials Science and Engineering, Tsinghua University, China) for conducting X-ray micro-computed tomography experiments.

References

[1] Terrani K A 2018 Accident tolerant fuel cladding development: Promise, status, and challenges *J. Nucl. Mater.* **501** 13–30

[2] Zhang X, Pan X, Lu Y and Zhang R 2022 Preparation and research progress of accident tolerant fuel pellets *Powder Metall. Technol.* **40** 334-339,350

[3] Ko J, Kim J W, Min H W, Kim Y and Yoon Y S 2022 Review of manufacturing technologies for coated accident tolerant fuel cladding *J. Nucl. Mater.* **561** 153562

[4] Snead L L, Venneri F, Kim Y, Terrani K A, Tulenko J E, Forsberg C W, Peterson P F and Lahoda E J 2011 Fully Ceramic Microencapsulated Fuels: A Transformational Technology for Present and Next Generation Reactors *Trans. Am. Nucl. Soc.* **104**

[5] Sonat Sen R, Boer B, Bess J D, Pope M A, and Ougouag A M 2012 *Advanced Fuels for LWRs: Fully-Ceramic Microencapsulated and Related Concepts FY 2012 Interim Report*

[6] Terrani K A, Snead L L and Gehin J C 2012 Microencapsulated fuel technology for commercial light water and advanced reactor application *J. Nucl. Mater.* **427** 209–24

[7] Terrani K A, Kiggans J O, Silva C M, Shih C, Katoh Y and Snead L L 2015 Progress on matrix SiC processing and properties for fully ceramic microencapsulated fuel form *J. Nucl. Mater.* **457** 9–17

[8] Olander D 2009 Nuclear fuels – Present and future *J. Nucl. Mater.* **389** 1–22

[9] Bragg-Sitton S 2014 Development of advanced accident-tolerant fuels for commercial LWRs *Nucl. News* **57** 83–91

[10] Zhou X, Yang Y, Song J, Lu Z, Zhang J, Liu B and Tang Y 2018 Carbon materials in a high temperature gas-cooled reactor pebble-bed module *New Carbon Mater.* **33** 97–108

[11] Zhang C, Wang Y, Wu Y, Liu S, Chen P, He Y, Yin C, Zhang J, Tian W, Su G H, Qiu S and Wu J 2021 Preliminary numerical investigation of TRISO-matrix interface debonding characteristics in fully ceramic microencapsulated fuel *Ann. Nucl. Energy* **159** 108338

[12] Terrani K A, Kiggans J O, Katoh Y, Shimoda K, Montgomery F C, Armstrong B L, Parish C M, Hinoki T, Hunn J D and Snead L L 2012 Fabrication and characterization of fully ceramic microencapsulated fuels *J. Nucl. Mater.* **426** 268–76

[13] Lian C 2019 *Fabrication of dispersion nuclear fuel with a highly uniform nuclide distribution by 3D printing* (China: University of Science and Technology of China)

[14] Lian C, Zhenqi C, Yongwen S, Lin G and Yuan W 2021 Fabrication of Dispersion Ceramic Nuclear Fuel by a Stereolithography-based Addictive Manufacturing Process *J. Mater. Sci. Eng.* **39** 9-15,57

[15] Ang C, Singh G, Snead L and Katoh Y 2019 Preliminary study of sintering zero-rupture Fully Ceramic Microencapsulated (FCM) fuel *Int. J. Appl. Ceram. Technol.* **16** 1699–707

[16] Ang C, Snead L and Kato Y 2020 A logical approach for zero-rupture Fully Ceramic Microencapsulated (FCM) fuels via pressure-assisted sintering route *J. Nucl. Mater.* **531** 151987

[17] Liu C, Huang R, Tan J, Lin H-T, Liu M, Liu B and Liu R 2022 Fully ceramic microencapsulated fuels fabricated by tape casting *J. Nucl. Mater.* **564** 153675

[18] Liu W, Shao Z, Liu W, Meng Y, Feng S and Cai Z 2023 Study on performance of coated particle dispersed fuel pellet prepared by spark plasma sintering *At. Energy Sci. Technol.* **57** 1810–6

[19] Snead L L, Terrani K A, Katoh Y, Silva C, Leonard K J and Perez-Bergquist A G 2014 Stability of SiC-matrix microencapsulated fuel constituents at relevant LWR conditions *J. Nucl. Mater.* **448** 389–98

[20] Cao F, Fan X, Liu B, Zhao X, Guo F and Xiao P 2018 Microstructure and thermal conductivity of fully ceramic microencapsulated fuel fabricated by spark plasma sintering *J. Am. Ceram. Soc.* **101** 4224–36

[21] Wang G, Lu G, Pei B and Wang G 1998 Oxidation mechanism of Si_3N_4-bonded SiC ceramics by CO, CO_2 and steam *J. Mater. Sci.* **33** 1309–17

[22] Gomez E, Echeberria J, Iturriza I and Castro F 2004 Liquid phase sintering of SiC with additions of Y_2O_3, Al_2O_3 and SiO_2 *J. Eur. Ceram. Soc.* **24** 2895–903

[23] Kim H, Kim Y and Lim K 2019 Pressureless sintered silicon carbide matrix with a new quaternary additive for fully ceramic microencapsulated fuels *J. Eur. Ceram. Soc.* **39** 3971–80

[24] Lu Z and Liu T 2018 Numerical simulation analysis of effective thermal conductivity of inert matrix dispersion pellet *Nucl. Sci. Eng.* **38** 710–7

[25] Zhang Y, He X, Qu X, Kong X and Duan B 2008 Densification process of spark plasma sintering for silicon carbide ceramic *Mater. Mech. Eng.* **32** 45–51

A Surface Composition Study of Niobium after BCP for SRF Cavity in Particle Accelerator

Didi Luo[1], **Changlin Wang**[1,*], **Feng Pan**[2,**]
[1] Institute of Modern Physics, Chinese Academy of Sciences, Nanchang Rd 509, Lanzhou, 730000, China
[2] GLVAC Industrial Technology Research Institute of High Power Devices Co. LTD, Suzhou, 215300, China

E-mail: *wangchanglin@impcas.ac.cn

E-mail: **panfeng@glvac.cn

Abstract. Nitrogen contamination from post-processing significantly affects the performance of niobium superconducting radiofrequency (SRF) cavities. Therefore, several performance-boosting strategies, including nitrogen doping, infusion and chemical acid reactions, have been explored in various studies. Among these studies, some secondary ion mass spectrometry (SIMS) results showed that nitrogen contamination exists on the niobium surface after the buffered chemical polishing (BCP) process. However, the compositions of these nitrogen byproducts were hard to detect. In this work, we calculated the chemical reaction process by the density function theory (DFT), predicted the reaction product, and then analyzed the Nb sample surface after BCP with X-ray Photoelectron Spectroscopy (XPS) to determine the composition of these contaminations.

1. Introduction

The Accelerator Driven Sub-critical System (ADS), which can transmute long-life radioactive waste, is considered as the optimal method of converting spent fuel into short-lived isotopes. As a concept for awaiting verification before widespread implementation, ADS is in the transition from the phase of developing key technologies to the systematic and integrated research.

As one national large-scale scientific instrumentation, the China initiative Accelerator Driven System (CiADS) [1] will be the world's first prototype of ADS facility at the megawatt level to explore the safe and proper nuclear waste disposal technology.

1.1. SRF Cavity

Particle accelerators have extensively adopted superconducting radiofrequency (SRF) cavities to achieve high accelerating gradients and low RF power dissipation [2]. The most frequently used measure of cavity performance is the Q_0 vs. E_{acc} plot (or the "Q vs. E curve"), where E_{acc} is the cavity acceleration gradient, and Q_0 is the intrinsic quality factor given by:

$$Q_0 = \frac{\omega U}{P_{loss}} = \frac{\Gamma}{R_S}, \tag{1}$$

where ω is the operation frequency, U is the storage energy, P_{loss} is the power dissipation, Γ is the cavity geometry factor, and R_S is the surface resistance.

Although their power losses are much lower than those of normal conducting RF cavities, superconducting cavities still suffer from multiple degrading effects that limit their performance, such as field emission (FE), multipacting, hydrogen Q-disease, Q-slope (HFQS) and thermal quench, resulting in both a lower Q than ideal theoretical prediction and a drop in Q as E_{acc} increases (i.e. Q-slope). The mechanisms behind some of these degrading effects are still unclear, and a summary can be found in Ref.[3].

1.2. BCP Reaction

Buffered chemical polishing (BCP) acid consists of nitric acid, hydrofluoric acid and phosphoric acid with a volume ratio of 1: 1: X (where X is a number between 1 and 4 that regulates the reaction rate). BCP is a commonly used treatment method for superconducting niobium cavities: it etches away about 100 μm pollution layer to obtain a clean Nb RF surface, thus reducing the RF loss of the superconducting cavity. After the BCP reaction, air exposure generates a native oxide layer ($\sim$ 5 nm, mostly Nb_2O_5) on the niobium surface. Moreover, Tuggle from JLAB has observed nitrogen by secondary ion mass spectrometry (SIMS) on BCP'ed niobium sample [4]. The RF surface is sensitive to surface impurities, which induce additional RF loss. For example, Ref. [5, 6] showed that oxygen vacancies in the oxide layer could be treated as magnetic impurities, and Ref. [7] showed that the Kapitza thermal boundary resistance generates ΔT heating. However, some side or intermediate reaction products, such as niobium nitride or oxynitride, may exist after BCP, and the study of RF dissipation from other byproducts still needs to be completed. Accordingly, studying the generated byproduct is necessary to evaluate the impact of the reaction process on the superconducting cavity's operation.

The overall reaction of BCP acid with niobium is: $Nb + 5\,HF + 5\,HNO_3 \longrightarrow NbF_5 + 5\,NO_2\uparrow + 5\,H_2O$, where the NbF_5 is soluble, and the NO_2 is gas. This work uses the first-principles calculations to explore the chemical reaction process by calculating the interaction between nitric acid and the niobium metal, and then verifies the results by experiments.

2. DFT Calculation

Based on Density functional theory (DFT), we use the siesta [8] program to optimize the architecture and calculate the system's stability. The exchange-correlation potential adopts the generalized gradient approximation (GGA) in the form of Perdew Burke Ernzerhof (PBE) [8]. In the calculation, we constructed a three-layer stack of niobium atoms and simulated the niobium surface by fixing the underlying atoms. In the Periodic boundary conditions, we use a unit cell with the size of 15 Å $\times$ 15 Å $\times$ 35 Å to simulate discrete HNO_3 molecules and the vacuum layer on the metal surface. Brillouin zone in inverted space adopted Monkhorst Pack grid centered on Γ point for sampling, using a k point grid of $2 \times 2 \times 1$. All lattice structures in this work were optimized using the conjugate gradient (CG) method [9], with a force convergence standard of 10^{-2} eV/Å.

2.1. Adsorption of Single HNO_3 Molecules on Nb Surface

The Nb (110) surface is the most stable, so we first optimized the structure of the Nb metal block and cut a thin layer structure containing three layers of atoms along the (110) surface to simulate the (110) surface of the Nb bulk. We also constructed a single HNO_3 molecular structure and optimized it. The optimized Nb (110) surface and HNO_3 molecular structure are shown in Fig.1. All HNO_3 atoms are in the same plane. To simulate the reaction process of HNO_3 molecules on the Nb (110) surface, we set up three initial configurations, allowing HNO_3 molecules to adsorb at varied positions on the Nb surface with distinct postures.

The first configuration is shown in Fig.2, where the HNO_3 molecule lies flat on the Nb surface and the N atom is next to a Nb atom. After this initial structural optimization, the HNO_3 molecule decomposes and forms O, OH, and NO atoms or groups that are adsorbed

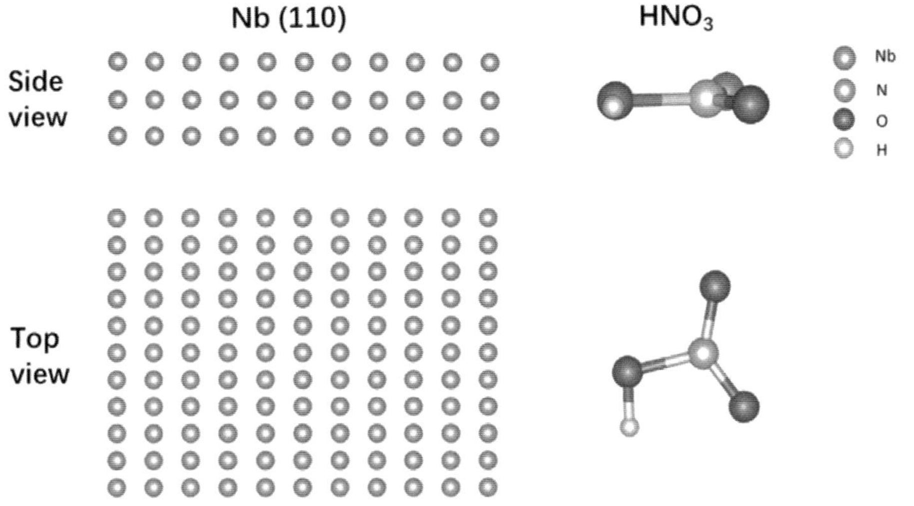

Figure 1. Schematic diagram of the structure of Nb (110) surface (left) and isolated HNO_3 molecule (right)

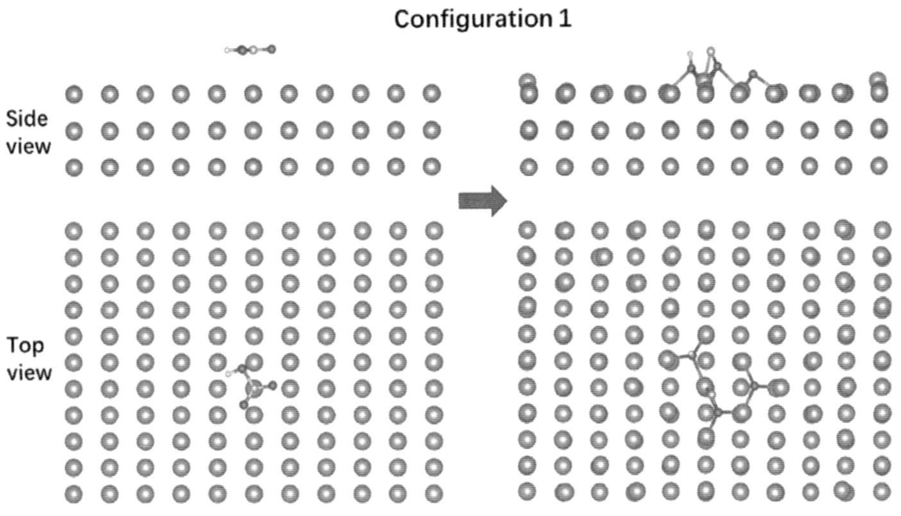

Figure 2. Schematic diagram of the structure of a single HNO_3 molecule adsorbed on Nb surface configuration 1 before and after optimization.

on the Nb surface, respectively, and are all adsorbed on the metal surface in the form of O atoms bonding with three neighboring Nb atoms. The calculated adsorption energy of HNO_3 molecule corresponding to the optimized structure is 6.822 eV (defined as the energy of the pure Nb (110) surface plus the energy of isolated HNO_3 molecules, minus the total energy of the optimized structure after HNO_3 adsorption on the Nb surface), which is the highest of the three configurations. This reaction generates oxides and oxynitrides.

The second configuration is shown in Fig.3, where the HNO_3 molecule remains flat on the Nb surface. Unlike configuration 1, two O atoms are next to the surface Nb atoms. After structural optimization, HNO_3 molecules still decompose, forming OH and NO_2 groups adsorbed on the

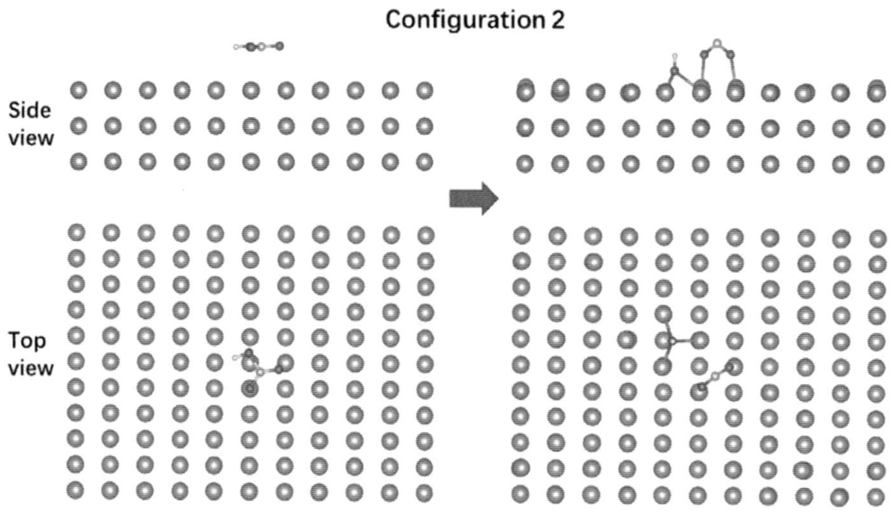

Figure 3. Schematic diagram of the structure of a single HNO_3 molecule adsorbed on Nb surface configuration 2 before and after optimization.

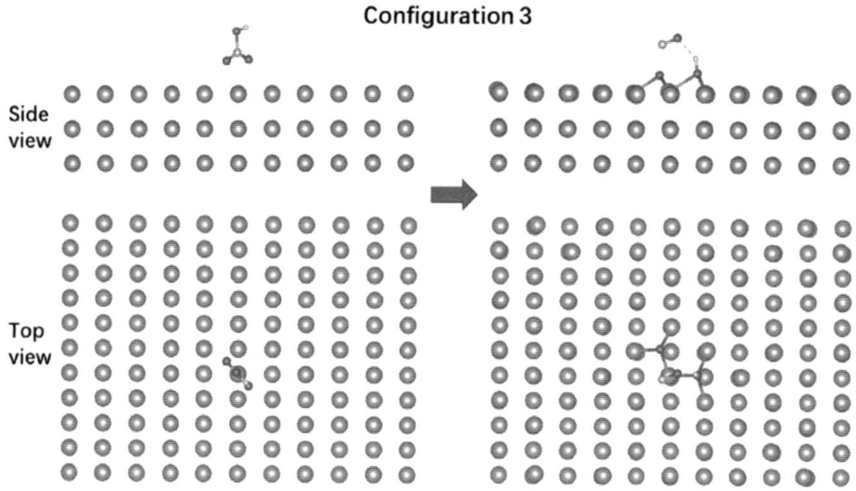

Figure 4. Schematic diagram of the structure of a single HNO_3 molecule adsorbed on Nb surface configuration 3 before and after optimization.

surface of Nb, respectively. OH is connected to three adjacent Nb atoms through O atoms, while NO_2 is adsorbed onto the metal surface through two O atoms connected to one Nb atom, respectively. This structure's adsorption energy is 4.985 eV, which is the lowest of the three configurations. This reaction generates oxynitrides.

The third configuration is shown in Fig.4, where the HNO_3 molecule is initially perpendicular to the Nb surface with two O atoms next to the metal surface. After optimization, one NO molecule detached from the metal surface, and O and OH were adsorbed on the surface of Nb, also through the bonding between O atoms and Nb atoms. For this case, the adsorption energy is 5.499 eV, between configurations 1 and 2. This reaction generates oxides.

From these results, we find that HNO_3 molecule always decomposes into small groups to

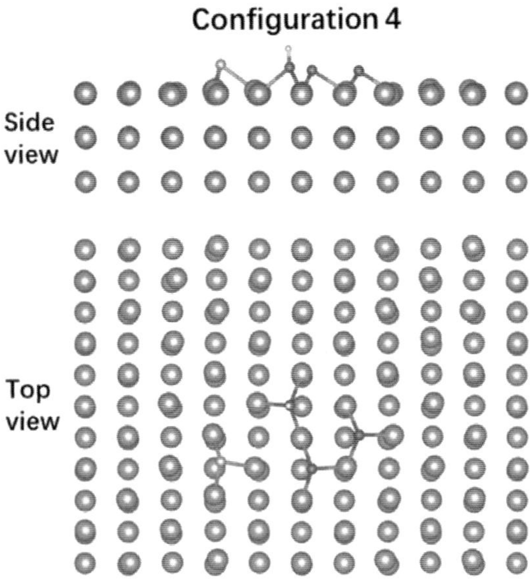

Figure 5. Schematic diagram of the structure of a single HNO_3 molecule adsorbed on Nb surface configuration 4 after optimization.

adsorb on the Nb (110) surface. The atoms and groups obtained from the decomposition of the most stable configuration 1 (i.e. the structure with the highest adsorption energy) are adsorbed on the Nb surface in a more dispersed form. The HNO_3 molecule is more inclined to bond with surface metal atoms to obtain a more stable structure.

We then constructed a configuration 4 structure to disperse and adsorb the N and O atoms in the HNO_3 molecule onto the Nb surface as much as possible. As shown in Fig.5, after optimization, the N atom and three O atoms all bond with three neighbouring Nb atoms, forming dispersed N, O, and OH. The calculated adsorption energy is 11.401 eV, significantly higher than the first three configurations. This high adsorption energy indicates that configuration 4 is the most stable case if the situation exists; however, this stable state may need to cross an energy barrier. This reaction generates oxides and nitrides.

2.2. Adsorption of Double HNO_3 Molecules on Nb Surface

To simulate the reaction of concentrated nitric acid on the Nb surface, we further calculated the adsorption of two HNO_3 molecules on the Nb surface considering the interaction between HNO_3 molecules. As shown in Fig.6, we placed two HNO_3 molecules connected by hydrogen bonds between H and O that lie flat on the surface of Nb. After optimization, HNO_3 molecules were decomposed into NO_2, NO, H_2O and O that adsorbed on the metal surface. The adsorption energy is 9.867 eV. The results also show that H and OH between adjacent HNO_3 molecules easily form water molecules, and this reaction still generates oxides and oxynitrides.

After optimization, the NO_2, NO, and O all bond with multiple Nb atoms, while H_2O only bonds with one Nb atom through the O atom, resulting in the easy departure of the H_2O molecule. Thus, we constructed two configurations after removing water molecules, as shown in Fig.7. Configuration 2 is optimized by removing H_2O from configuration 1, while configuration 3 disperses N and O atoms as much as possible on the Nb surface, yielding corresponding adsorption energies of 11.653 eV and 22.015 eV, respectively. These values show that the energy of N and O atoms adsorbing on the Nb surface is much lower than NO or NO_2 groups. For

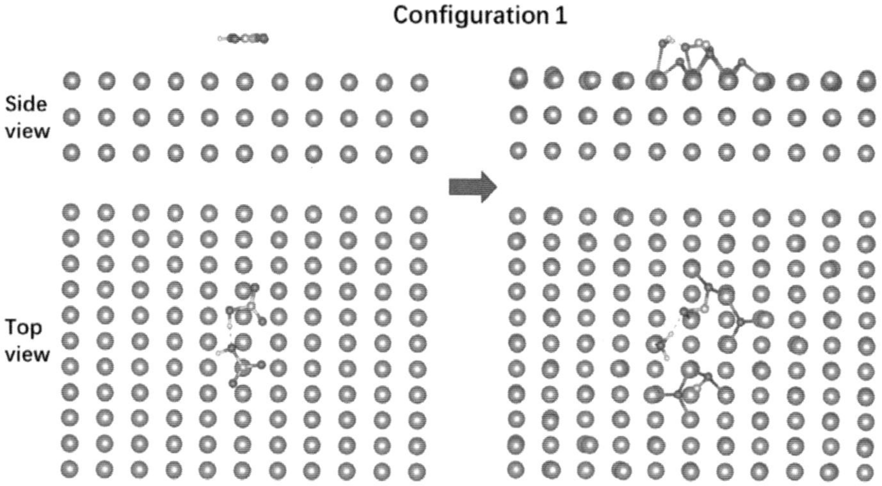

Figure 6. Schematic diagram of the structure of two HNO_3 molecules adsorbed on Nb surface configuration 1 before and after optimization.

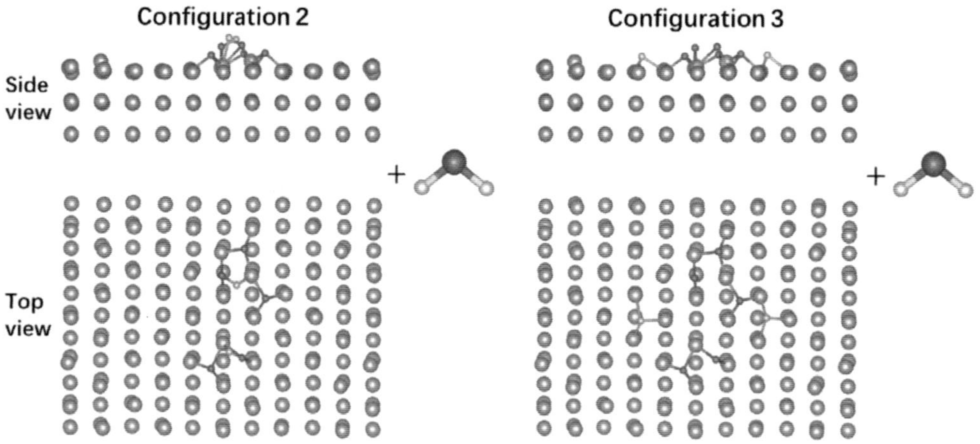

Figure 7. Schematic diagram of the structure of two HNO_3 molecules adsorbed on Nb surface configuration 2 and 3 after optimization.

configuration 3, the surface may generate oxides and nitrides when above specific temperatures.

Subsequently, we investigated the potential barriers for forming oxides and nitrides from nitrogen oxide groups. As shown in Fig.8, we take configuration 1 directly optimized by two HNO_3 molecules as the initial structure, and move the N atoms not bonded to the metal surface to the nearby Nb atoms to form the dispersed adsorbing atoms as the target structure (i.e. configuration 4) and calculate the transition state structure from configuration 1 to configuration 4 and the corresponding energy barrier.

The primary transformation from configuration 1 to configuration 4 is that both N atoms break the bond with the O atoms and connect to the Nb atoms. The transition state structure is similar to configuration 1. During the exploration of the transition state, we found that when the structure transforms from configuration 1 to configuration 4, a slight structural change will lead to the system's total energy decrease, only needing to cross a shallow barrier (ΔE = 53

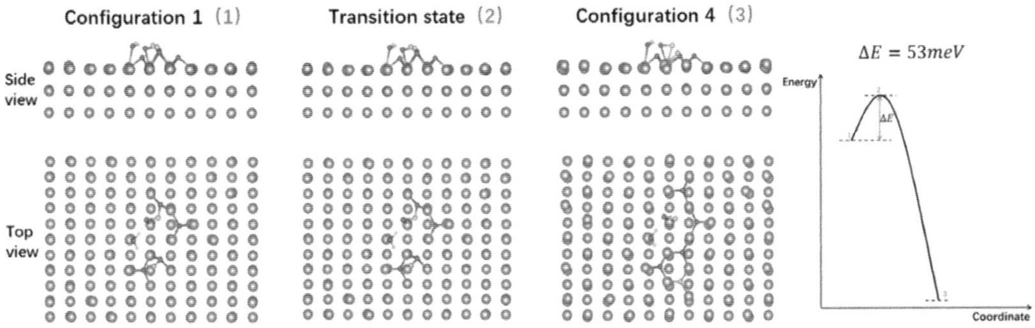

Figure 8. Schematic diagram of the structural change and energy barrier required for the adsorption of double HNO_3 molecules on the Nb surface from configuration 1 through a transition state to configuration 4. The structures represented by the red numbers 1, 2 and 3 in the energy barrier graph (right 1) correspond to the left configuration 1, transition state and configuration 4 structures, respectively.

meV). The transitions from configuration 1 result to configuration 4 decrease the total energy by 7.625 eV, indicating significantly higher stability. Once the system crosses the barrier, the surface nitrides will be very stable.

3. Experiments

We ordered the RRR = 300 Nb sheets from NIN [10]; both sheets were cold rolled to 1 mm and annealed at 700 °C for 10 hours. We performed BCP on the samples for about 100 μm, and then did XPS analysis to study the niobium surface components.

3.1. Bulk Nb Reacts with BCP 100 μm: Same as Cavity Process

To characterize the composition in detail, we performed X-ray Photoelectron Spectroscopy (XPS) on the Nb sample. The equipment is Thermo Scientific K-Alpha+ from ThermoFisher, the vacuum in the chamber is about 5×10^{-9} mbar, the X-Ray source is Mono AlKa (1486.6 eV, 12 KV, 6 mA), the scan mode is CAE, and the work function is 4.2 eV.

Fig.9 shows the XPS survey of bulk niobium reaction with BCP for 100 μm (a). This result shows only O, C and Nb signals. Here, carbon absorption is normal contamination for XPS results. Tuggle has observed nitrogen by SIMS on BCP'ed niobium sample [4]: nitrogen element mostly stays on the surface 0.05 μm and only has a very trace amount, making the detection very difficult. For this BCP sample, we have one hypothesis that the niobium surface roughness after 100 μm BCP reaction is μm-level ($R_p \sim 5\mu m$), resulting in difficult XPS signal detection for the tiny amount of nitrogen. Hence, we need to rule out the high roughness influence. One possible way is to coat niobium on a very smooth substrate that does not react with BCP at the reaction temperature (room temperature). Niobium films growing on Al_2O_3 substrate were selected as the target.

3.2. BCP Light Reaction (0.3 μm) with 2 μm Nb Coating on Al_2O_3 Sample

In order to reduce the roughness impact, we performed niobium magnetron sputtering on an alumina substrate to obtain $\sim$ 2 μm niobium film. We then put the sample into BCP acid for a short period (20 seconds) and washed it with ultra-pure water. The reaction etched away about 0.3 μm Nb, and roughness should be less than 1 μm after the reaction. We later performed XPS to characterize the surface component and Nb oxidation state, as shown in Fig.9 (b). We also

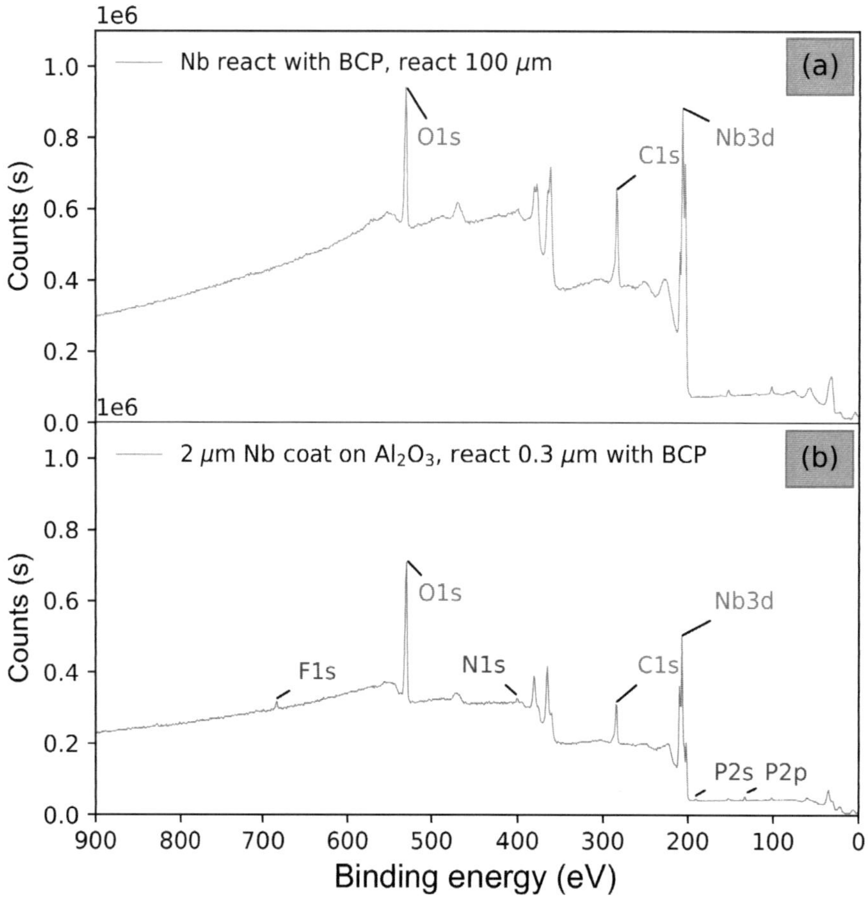

Figure 9. XPS survey of the bulk niobium react with BCP for 100 μm (a), and the 2 μm niobium coating on Al_2O_3 and react with BCP for 0.3 μm (b). The characteristic peaks are marked in the corresponding positions.

performed Ar^+ sputtering to obtain the sample composition profiling, as shown in Fig.10. We found the listed information:

(i) Unlike the baseline BCP 100 μm result, some weak F, N, and P peaks emerge from the Nb film sample XPS (Fig.9), indicating this strategy improved the test sensitivity for low concentration elements.

(ii) Fig.10 shows that N and P disappeared after only 7 seconds Ar^+ sputtering, indicating the N and P elements exist only in the first 0.6 nm, agreed with the results in Ref.[4].

(iii) The F1s signal stays for all the sputtering results. Even if HF does not react with Al_2O_3 below 1000 °C, it may still easily be absorbed by the Al_2O_3 surface and influence the sample XPS result.

(iv) The O1s signals stay the same during the 35 seconds of sputtering, indicating the Nb film is highly oxidized in the first 3 nm. This oxidization could be from the BCP reaction or from the Nb film coating process.

Thus, we conclude that the Nb film reacting with the BCP strategy has too many uncertainties, so we chose another way to eliminate the high-roughness parameter of the BCP sample.

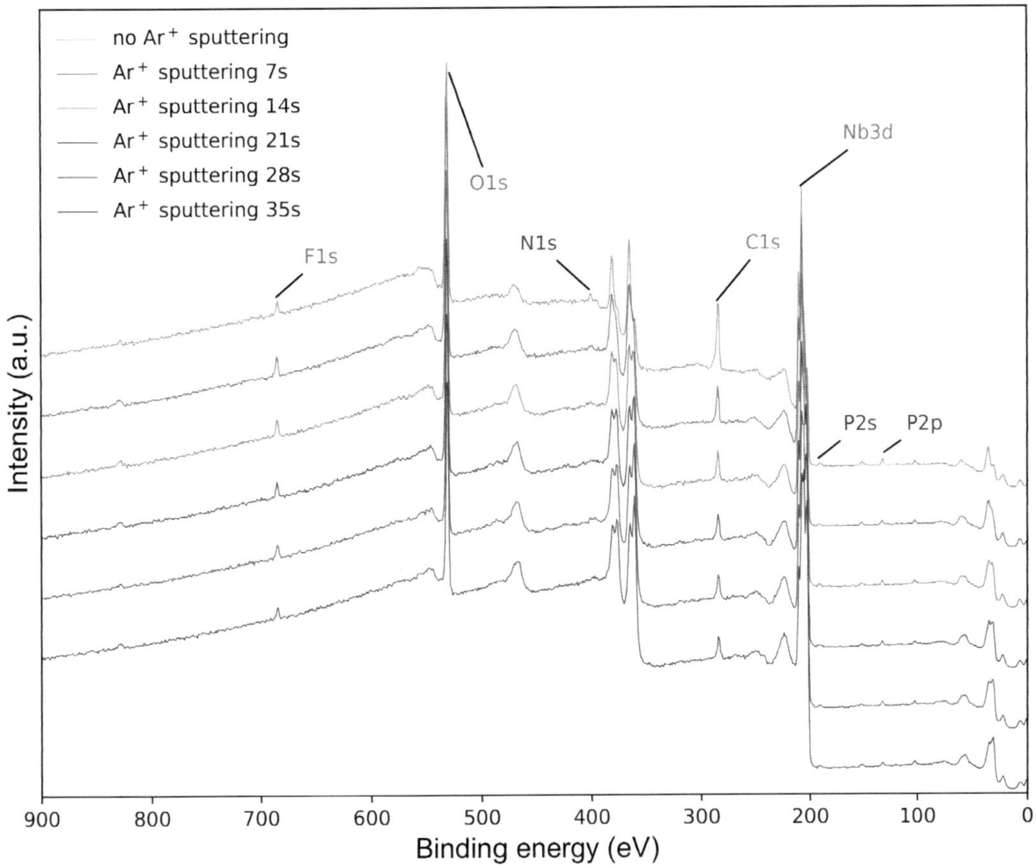

Figure 10. The XPS signal of Ar^+ sputtering process for the 2 μm niobium coating on Al_2O_3 and react with BCP for 0.3 μm sample. The Ar^+ sputtering rate is about 4.8 nm/min [11].

3.3. BCP Light Reaction (1 μm) with Mechanical Polished Nb Sample

We performed mechanical polishing for a bulk niobium sample, obtained a nanometer-level polished surface, and let the sample react with BCP acid for one minute (etched about 1 μm), rinsed with ultra-pure water. In this case, the roughness should still be less than 1 μm. We then characterized the surface by XPS (Fig.11 (c)). We also performed Ar^+ sputtering to obtain the sample composition change with depth, as in Fig.11 (d). From the test, we found:

(i) Some F, N, and P signals exist on the 1-μm-reaction sample; the N and P peaks are even higher than the Nb film sample, indicating this strategy improved the test sensitivity for low-concentration elements.

(ii) The O1s signals reduced during the 34 seconds of sputtering, indicating that the Nb oxides are mostly located in the first 3 nm. Unlike the film-coating on Al_2O_3 sample, the mechanical polishing does not oxidize the high-purity bulk niobium sample in the deep.

3.4. Preliminary Results of the N form on Nb surface

We thus acquired the high-resolution signals of the N1s (Fig.12), O1s (Fig.13) and Nb3d (Fig.14) to study the oxidation state of niobium. For N1s, the N-O dominates the signal, indicating that configuration 1 in Fig.6 or configuration 2 in Fig.3 results happen, which means the NO or NO_2

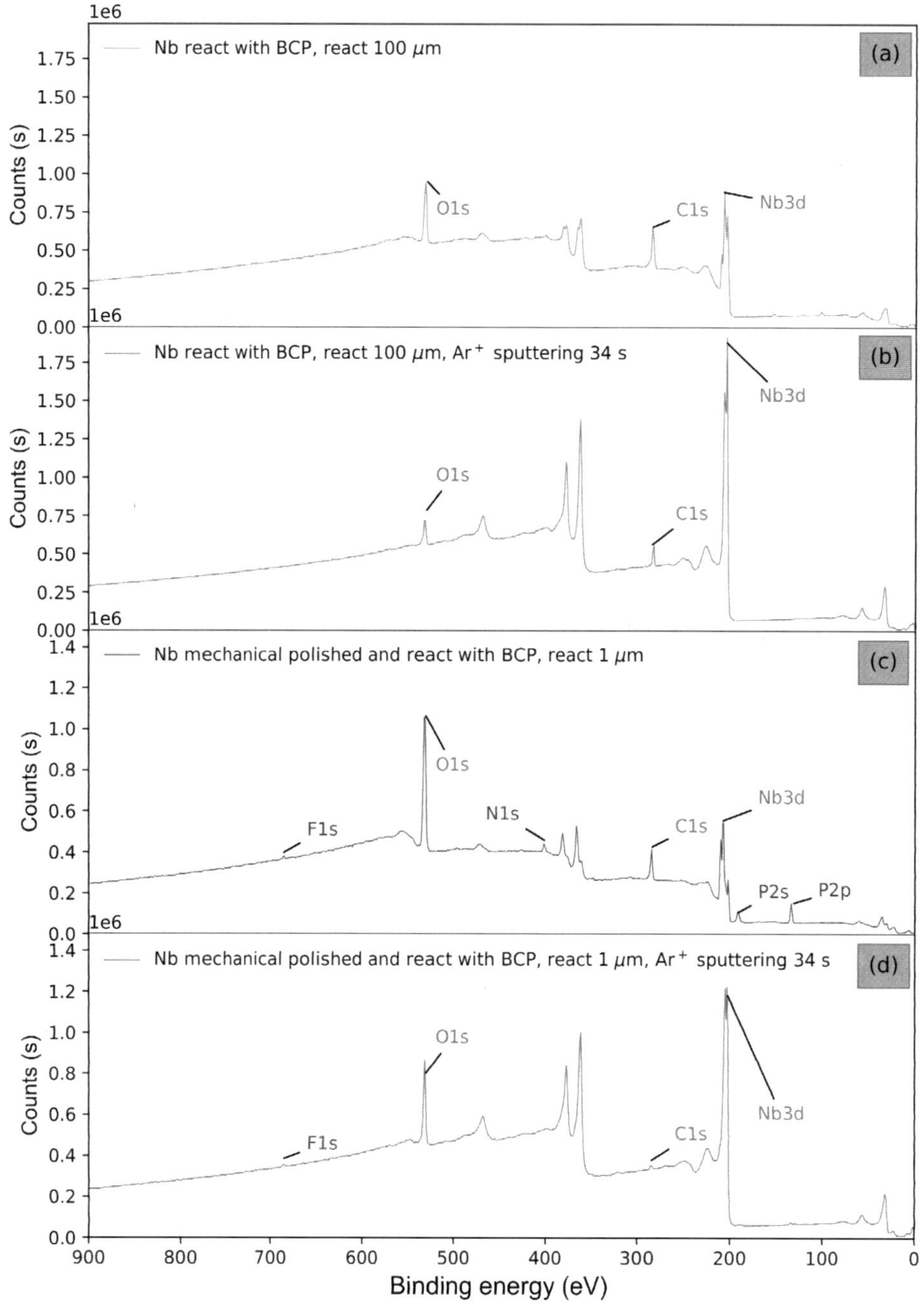

Figure 11. XPS survey of the bulk niobium react with BCP for 100 μm (a & b), and the mechanically polished bulk niobium react with BCP for 1 μm (c & d). Please notice the y-axis are different for the two samples. The Ar^+ sputtering rate is about 4.8 nm/min [11]).

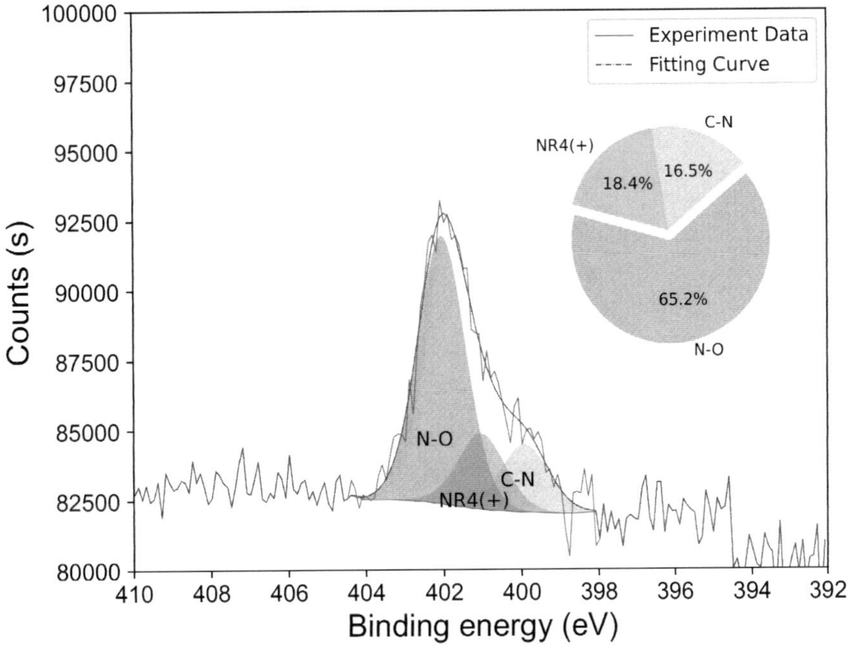

Figure 12. High-resolution XPS scans of N1s region for the niobium sample.

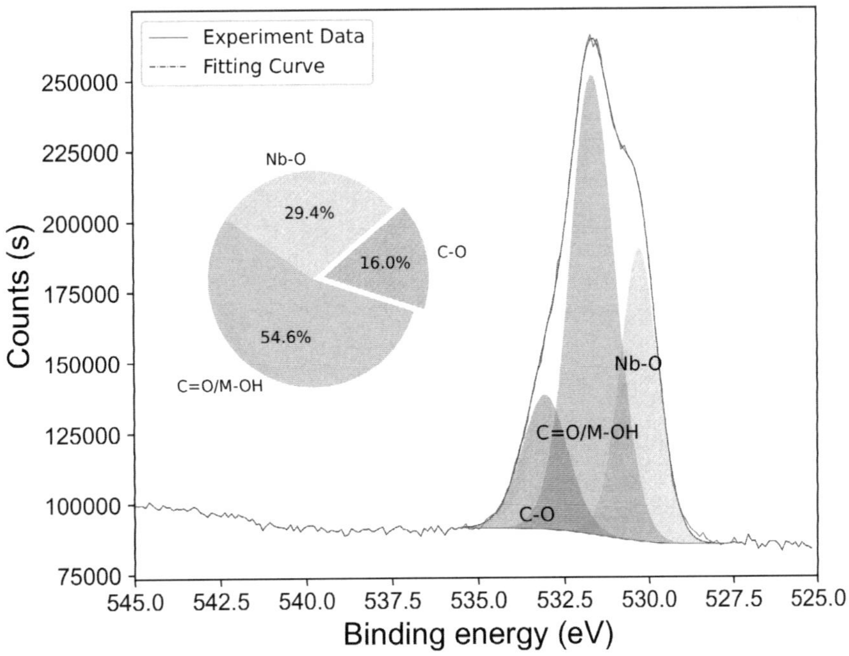

Figure 13. High-resolution XPS scans of O1s region for the niobium sample. The valence state of Nb-O is Nb^{2+}.

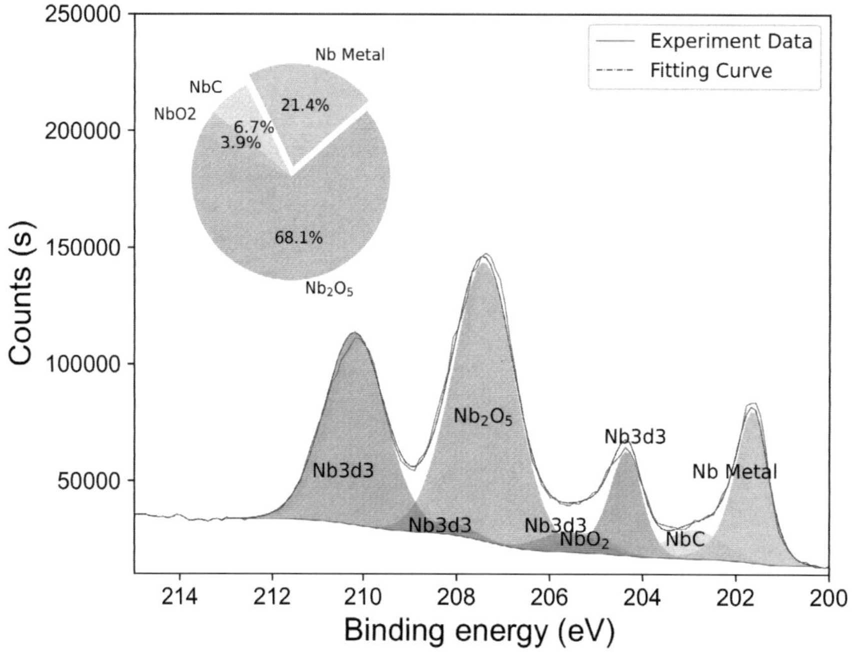

Figure 14. High-resolution XPS scans of Nb3d region for the niobium sample. The Nb3d3 signals are filled grey. The valence state of Nb is Nb^{2+} in NbC, Nb^{4+} in NbO_2 and Nb^{5+} in Nb_2O_5, respectively.

group absorbed on the niobium surface. Fig.14 also shows the absence of Nb-N bonding. The XPS of N1s and Nb3d high-resolution signal has no niobium nitride composition, indicating that configuration 4 in Fig.5 does not exist. The barrier $\Delta E = 53$ meV calculated from Fig.7 and Fig.8 is not crossed. Fig.13 shows Nb-O bonding and carbon contamination, maybe due to the low concentration of the NO group.

4. Conclusion

In summary, we used the first-principles calculation to simulate the interaction between a single and two HNO_3 molecules in contact with the surface of Nb metal (110). Calculations indicated that HNO_3 molecules are easily decomposed into OH, NO, and NO_2 groups adsorbed on the surface when in contact with Nb. The reaction can generate oxides and oxynitrides (configuration 1, 2 & 3 for single HNO_3 absorption, configuration 1 & 2 for double HNO_3 absorption) or oxides and nitrides (configuration 4 for single HNO_3 absorption, configuration 3 & 4 for double HNO_3 absorption). Further calculation showed that nitride formation needs to cross a $\Delta E = 53$ meV barrier.

We then did experiments to explore the byproducts of the BCP reaction. We first let the Nb bulk sample react with BCP for 100 μm, the same as the cavity treatment, and did not find the N signal in the XPS. After improving the niobium surface roughness by coating niobium on an alumina substrate and mechanically polishing with a light BCP reaction, we successfully obtained N, F, and P peaks on the niobium surface. High-resolution XPS scans of N1s, O1s and Nb3d only showed the existence of N-O bonding, indicating the BCP reaction at room temperature did not cross the 53 meV barrier.

In the future, we will further optimize the surface condition to get higher N signals, explore the F and P interactions with the niobium surface, and get a complete picture of BCP reaction byproducts. This strategy can also be applied to characterize the chemical reactions of the prevalent nitrogen doping and infusion techniques.

Acknowledgments
The work was supported by the Young Scientists Fund of the National Natural Science Foundation of China (Grant NO. 12105335).

References

[1] Ciads, Website, https://english.imp.cas.cn/research/facilities/CIADS/ (2022).
[2] M. Tigner, Rf superconductivity for accelerators–is it a hollow promise?, IEEE Transactions on Magnetics 15 (1979) 15–20.
[3] H. Padamsee, History of gradient advances in srf (2020). arXiv:2004.06720.
[4] J. Tuggle, A. Palczewski, C. Reece, F. Stevie, M. Kelley, Investigation of low-level nitrogen in niobium by secondary ion mass spectrometry, in: Proceedings of LINAC2016, Virginia Polytechnic Institute and State University, East Lansing, MI, USA, 2016.
[5] T. Proslier, J. Zasadzinski, L. Cooley, C. Antoine, J. Moore, J. Norem, M. Pellin, K. Gray, Tunneling study of cavity grade nb: Possible magnetic scattering at the surface, Applied Physics Letters 92 (04 2009). doi:10.1063/1.2913764.
[6] M. Wenskat, J. Čižek, M. O. Liedke, M. Butterling, M. Stiehl, G. D. L. Semione, C. Backes, C. Bate, O. Melikhova, E. Hirschmann, A. Wagner, H. Weise, A. Stierle, M. Aeschlimann, W. Hillert, Vacancy dynamics in niobium and its native oxides and their potential implications for quantum computing and superconducting accelerators, Phys. Rev. B 106 (2022) 094516. doi:10.1103/PhysRevB.106.094516.
URL https://link.aps.org/doi/10.1103/PhysRevB.106.094516
[7] J. D. Halbritter, Degradation of superconducting rf cavity performances by extrinsic properties, 11th Workshop on RF-Superconductivity (SRF 2003), Lübeck, September 8-12, 2003 Proc.on CD-ROM MoP44 Hamburg : DESY, 2004, 42.02.02; LK 01 (2005).
[8] J. M. Soler, E. Artacho, J. D. Gale, A. García, J. Junquera, P. Ordejón, D. Sánchez-Portal, The siesta method for ab initio order-n materials simulation, Journal of Physics: Condensed Matter 14 (11) (2002) 2745. doi:10.1088/0953-8984/14/11/302.
URL https://dx.doi.org/10.1088/0953-8984/14/11/302
[9] M. R. Hestenes, E. Stiefel, Methods of conjugate gradients for solving linear systems, Journal of research of the National Bureau of Standards 49 (1952) 409–435.
[10] Nin, Website, http://www.c-nin.com/ (2022).
[11] M. Yu, G. Pu, Y. Xue, S. Wang, S. Chen, Y. Wang, L. Yang, Z. Wang, T. Zhu, T. Tan, Y. He, S. Huang, K. Zhang, The oxidation behaviors of high-purity niobium for superconducting radio-frequency cavity application in vacuum heat treatment, Vacuum 203 (2022) 111258. doi:https://doi.org/10.1016/j.vacuum.2022.111258.
URL https://www.sciencedirect.com/science/article/pii/S0042207X22003827

High-Performance PVDF-HFP-PVAc-LLZTO Composite Solid-State Electrolyte for Lithium Metal Battery

Kaifeng Wang[+], Linghao Zhang[+], Ruoying Wu, Xin Xu, Lai Wei[*], Yunfang Gao[*]

College of Chemical Engineering, Zhejiang University of Technology, Hangzhou, Zhejiang 310014, P. R. China

[+]K. Wang and L. Zhang contributed equally to this work.

[*]Corresponding Author

E-mail address: 2469046108@qq.com (Lai Wei); gaoyf@zjut.edu.cn (Yunfang Gao)

Abstract: Composite solid-state electrolytes are highly promising contenders for next-generation lithium-metal batteries due to their high safety, good flexibility, and compatibility with lithium metal. In this work, Poly(vinyl acetate) (PVAc) was incorporated into PVDF-HFP in order to reduce the crystallinity of the PVDF-HFP matrix, improve the ionic conductivity of the electrolyte, and greatly reduce the interfacial impedance of the electrolyte by utilizing the low-temperature flexibility, excellent amorphous nature, and strong bonding properties of PVAc. On this basis, the active filler LLZTO was introduced to further reduce the crystallinity of the PVDF-HFP matrix and provide more Li^+ transport channels. The PP60-10LLZTO composite solid electrolyte has exceptional mechanical and electrochemical characteristics with a high ionic conductivity of $7.2*10^{-4}$ S cm^{-1} and the Li^+ transference number of 0.57 at 30 °C. The assembled LFP/Li battery can be stably cycled for 450 cycles at 30 °C and 0.5C, and its capacity retention rate is as high as 89%. The strategy of constructing composite solid-state electrolytes in this paper presents new insights for designing and developing high-performance solid-state lithium metal batteries.

Keywords: Solid state electrolyte, Lithium metal battery, PVDF, LLZTO

1. Introduction

Lithium-ion batteries are rapidly gaining prominence in the market for small electronic devices due to their diminutive size, light weight, high operating voltage and substantial energy density [1,2]. Nevertheless, their application in the automotive industry and large-scale energy storage requires further improvements in factors such as energy density, durability and safety [3,4]. The substitution of liquid electrolytes with polymer solid electrolytes is considered a crucial solution to address these challenges. Polymer solid-state electrolytes act as both a diaphragm and an electrolyte, as well as a physical barrier to lithium dendrite growth, which fundamentally resolves the issues of low energy density and poor safety associated with lithium-ion batteries [5-7]. However, polymer solid-state electrolytes have some inherent disadvantages, including suboptimal ionic conductivity (IC) at ambient temperature, diminished Li^+ mobility, and inadequate mechanical characteristics [8, 9].

To mitigate these issues, several modifications have been proposed, such as the incorporation of inorganic fillers, copolymerization, blending, and plasticisation [10-12]. The blending of polymers can amalgamate the advantages of different host polymers to synthesize a polymer matrix (PM) with excellent properties [9, 13, 14]. Polyvinyl acetate (PVAc), as an amorphous polymer with low-temperature flexibility, excellent amorphous properties, large dipole moment, and high dielectric constant, presents a viable solution. By incorporating PVDF-HFP, the flexibility of the matrix can be enhanced, interfacial impedance reduced, the crystallinity of the PM decreased, and an additional kinetic channel for ionic excursion provided.

Incorporating inorganic fillers is another effective strategy to enhance the electrochemical performance of PVDF-HFP. These fillers are mainly categorized as inert fillers (e.g., Al_2O_3, TiO_2, ZrO_2, etc.) or active fillers (e.g., $Li_{6.5}La_3Zr_{1.5}Ta_{0.5}O_{12}$ [15-17], $Li_{10}GeP_2S_{12}$ [18], etc.). Active fillers not only attenuate the crystallinity of polymer solid-state electrolytes but also offer more transport channels for lithium ions as an inorganic ceramic solid-state electrolyte itself, making them widely used in organic-inorganic composite solid-state electrolytes [19-21]. Zhang and co-workers introduced cubic-phase LLZTO solid state electrolytes into PVDF-based SPEs, inducing a chemical dehydrogen fluorination reaction in the PVDF skeleton. The alterations in PVDF chains induced a synergistic interplay among the LLZTO filler, PVDF matrix, and lithium salt, leading to a substantial enhancement in the performance of the composite solid-state electrolyte. As a result, the electrolyte exhibited noteworthy improvements, including an IC of approximately $5*10^{-4}$ S/cm at 25 °C, elevated mechanical strength, and favorable thermal stability.

This paper aims to improve the properties of PVDF-HFP by incorporating PVAc. By leveraging PVAc's low-temperature flexibility, excellent amorphous nature, and strong bonding properties, we could reduce the crystallinity of the PVDF-HFP matrix, enhance the IC of the electrolyte, and significantly decrease the interfacial impedance of the electrolyte. Consequently, the active filler LLZTO was used to enhance the PVDF-HFP matrix's Li^+ transport capabilities by reducing its crystallinity and constructing more transport channels. The prepared PP60-10LLZTO composite solid electrolyte exhibits good electrochemical and mechanical properties, with an IC of $7.2*10^{-4}$ S/cm at 30 °C and a Li^+ transference number of 0.57. The assembled LFP/Li battery is stable for 450 cycles at 30 °C and 0.5 C at constant current with a capacity retention of up to 89%. XPS and SEM analyses revealed that the composite polymer solid-state electrolyte could form a stable SEI film on the surface of the LA, suppress the growth of lithium dendrites, and prevent side reactions between the electrolyte and lithium metal.

2. Experimental section

2.1 Preparation of LLZTO

The conventional solid-phase reaction method was used to prepare $Li_{6.5}La_3Zr_{1.5}Ta_{0.5}O_{12}$. La_2O_3, LiOH, Ta_2O_5 and ZrO_2 were weighed according to stoichiometric ratios. Subsequently, 15 wt.% of LiOH was added to replenish the loss of Li during high-temperature sintering. The weighed raw materials were then added to isopropanol and ball-milled for 12 h. The ball-milled slurry was oven-dried for 12 h, and then sintered in a muffle furnace at 900 °C for 12 h to obtain the cubic-phase LLZTO. Subsequently, the cubic LLZTO powder was subjected to an additional 12-hour ball milling process for further refining LLZTO particles.

2.2 Preparation of solid-state electrolyte

PVDF-HFP (1.00 g), PVAc (0.30 g) and LiTFSI (0.78 g; 60% of the PM's mass) were added to 10 mL of N-N dimethylacetamide and stirred at a constant temperature of 60 °C for 6 h. Then, different mass ratios (5 wt.%, 10 wt.%, and 15 wt.%) of LLZTO powders (mass ratio of LLZTO to PM) were added to the above mixed solution, followed by continuous stirring at a constant temperature of 60 °C for 12

h and ultrasonication for a period of time to obtain a homogeneous brown precursor dispersion. The brown precursor dispersion was applied onto a pristine glass plate and levelled to a thickness of 500 μm using a spatula. Subsequently, it underwent drying under vacuum conditions. The resulting CSE was labeled as PP60-xLLZTO, where x represents the mass fraction of LLZTO to the PM. Furthermore, for comparative purposes, polymer solid electrolytes were prepared using PVDF-HFP and PVDF-HFP-PVAc based on the aforementioned procedure and designated as P60 and PP60, respectively. Subsequently, all the prepared solid electrolytes were cut into 19-mm diameter discs using a slicer.
The viscosity average molecular weight of the PVDF-HFP utilized in this experiment was 455,000, while that of PVAc was 500,000.

2.3 Electrochemical characterization

Symmetric cells consisting of SS/CPE/SS were assembled under a nominal pressure of 50 kPa. Electrochemical Impedance Spectroscopy (EIS) measurements were conducted over a frequency range of 10^{-2} Hz to 10^{6} Hz, spanning temperatures from 30 to 80 °C. The IC (σ) was determined through the computation of Equation (1):

$$\sigma=\frac{L}{R\times S} \quad (1)$$

where R (Ω) represents the impedance value, S (cm^2) denotes the area of the stainless-steel sheet, and L (cm) signifies the thickness of the composite solid electrolyte membrane. The Arrhenius behaviour of IC is determined through Equation (2):

$$\sigma=\sigma_0\exp(-\frac{Ea}{kT}) \quad (2)$$

where $\boldsymbol{\sigma_0}$ represents the prefactor, which is associated with the number of charges. **Ea** represents the activation energy of Li^+ conductivity, and **k** and **T** denote Boltzmann's constant and thermodynamic temperature, respectively. The electrochemical window of CPEs can be calculated by assembling SS/CPE/Li cells through linear scanning voltammetry (LSV; scan rate = 5 mV/s, voltage range = 2-7 V). The Li/CPE/Li symmetric cells were constructed according to Equation (3):

$$t^+=\frac{I_S\times(\Delta V-I_0\times R_0)}{I_0\times(\Delta V-I_S\times R_S)} \quad (3)$$

The Li^+ (**t^+**) transference number was determined, where **Δ V** is the applied 10 mv DC polarization voltage, **Is** and **I_0** represent the steady and initial state currents, while **Rs and R_0** denote the resistance before and after polarization. These measurements were conducted on an electrochemical workstation (CHI 760E, Shanghai Chenhua Instrument) under consistent temperature and voltage settings. The cycling performance of all-solid-state Li-ion batteries containing CPEs was examined on the Blue Power Test System CT2001A instrument. The tests were conducted in the voltage range of 2.5-4.0 V, with a current density of 1C (1C = 170 mAh/g), and temperature of 60 °C. The batteries underwent cycling performance assessments without additional stress during electrochemical testing.

2.4 Physical characterization

The sample morphology was analyzed using scanning electron microscopy (SEM, Gemini 500, Zeiss, Germany) and transmission electron microscopy (TEM, Tecnai G2 F30, FEI, The Netherlands). The crystalline phases were identified employing a Cu Kα radiation x-ray diffractometer (XRD, D8 Advance, BrukerAXS, Germany). The thermal stability of the Composite Solid Electrolytes (CPEs) was explored through thermogravimetric analysis (TGA, NETZSCH STA 449F3) and differential

Journal of Physics: Conference Series **2749** (2024) 012014 doi:10.1088/1742-6596/2749/1/012014

scanning calorimetry (DSC, NETZSCH DSC 200F3). Nitrogen absorption-desorption isotherms were acquired using a Micrometrics ASAP 2020 instrument.

3. Results and discussion

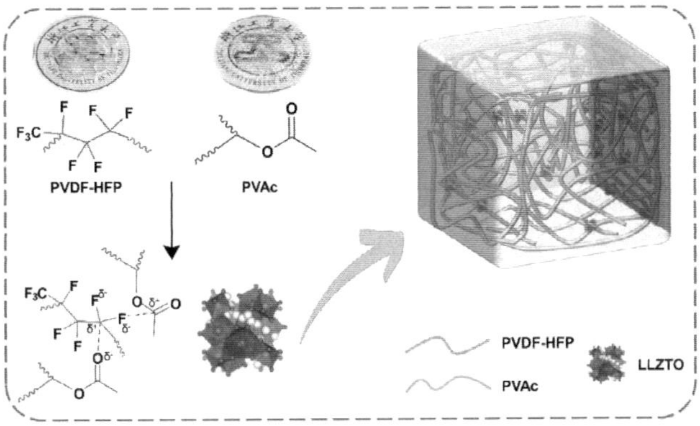

Figure 1. Schematic illustration for the synthesis mechanism of PP60-10LLZTO solid electrolyte.

Figure 1 depicts a schematic illustrating the synthesis mechanism of the PP60-10LLZTO solid-state electrolyte. In this illustration, the fluorine (F) in the C-F bond of PVDF-HFP attracts covalently bonded electrons, causing them to shift towards F due to its strong electronegativity. This process forms a C-positive intermediate, subsequently attracting lone-pair electrons in the C=O bond of PVAc towards PVDF-HFP. Additionally, F in PVDF-HFP attracts the C-positive intermediate in PVAc, disrupting the regular molecular ordering of PVDF-HFP and facilitating ion migration. Moreover, F in HFP attracts the C-positive intermediates in PVAc, further disrupting the regular molecular ordering of PVDF-HFP, thereby reducing the crystallinity of the PM and facilitating ion migration. Next, LLZTO active filler was prepared using the conventional solid-phase reaction method, and its XRD pattern is presented in Figure 2a. The diffraction peaks of LLZTO particles closely match the known standard card of garnet-type cubic phase Li5La5Nb2O12 (JCPDS File No. 80-0457), indicating a single-phase structure without any obvious heterogeneous peaks. The presence of Ta element doping leads to increased lithium vacancies, stabilizing the cubic phase structure, and resulting in uniformly cubic-phase products. Based on the SEM photographs (Figure 2b-d), it is evident that the LLZTO particles obtained during the preparation process are uniform in size, measuring approximately 1.5 μm. The incorporation of LLZTO active particles uniformly dispersed in the solid electrolyte contributes to the reduction of the PM's crystallinity, an increase in the amorphous region, and the creation of additional pathways for Li^+ transport. Simultaneously, the inclusion of LLZTO enhances the mechanical strength of the solid-state electrolyte, acting as a barrier against the infiltration of lithium dendrites and extending the service life of the solid-state battery.

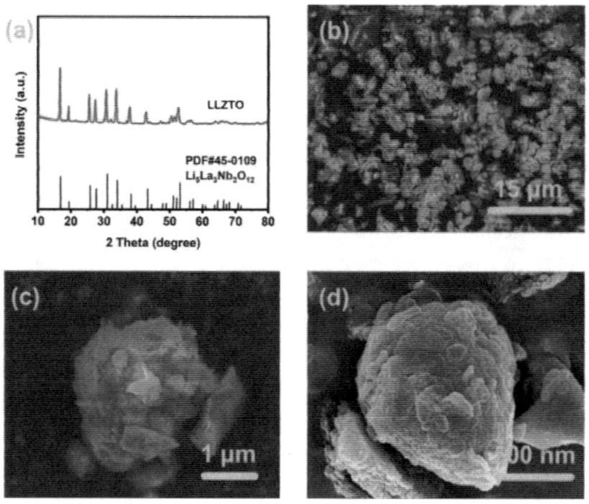

Figure 2. (a) XRD pattern and (b, c, d) SEM images of LLZTO.

Figure 3 presents SEM images of the P60, PP60, and PP60-10LLZTO electrolyte membranes. Figures 3a and 3b depict the SEM image of the P60 polymer solid electrolyte surface, composed only of PVDF-HFP and lithium salts. The images reveal a surface structure consisting of large interconnected spheres, with numerous pores between them. This configuration may not facilitate optimal contact between the electrolyte membrane and the electrode material. Figures 3c and 3d demonstrate the surface structure of the PP60 co-polymer solid electrolyte membrane after the addition of PVAc. It can be seen from the figures that with the addition of PVAc, the pore structure on the membrane surface was slightly improved and the particle size of the spherical structure was significantly reduced. This change may be attributed to increased system viscosity and a slower solvent evaporation rate due to the addition of PVAc. Figures 3e and 3f illustrate the PP60-10LLZTO composite polymer solid electrolyte film, prepared by further incorporation of LLZTO inorganic filler on top of PVAc. The addition of LLZTO leads to tighter interconnections between the spherical structures. Although a pore-like structure persists due to solvent evaporation, the number and size of the pores are notably reduced. This modification facilitates close contact between the electrolyte membrane and the electrode material, consequently enhancing the cycling performance of the battery.

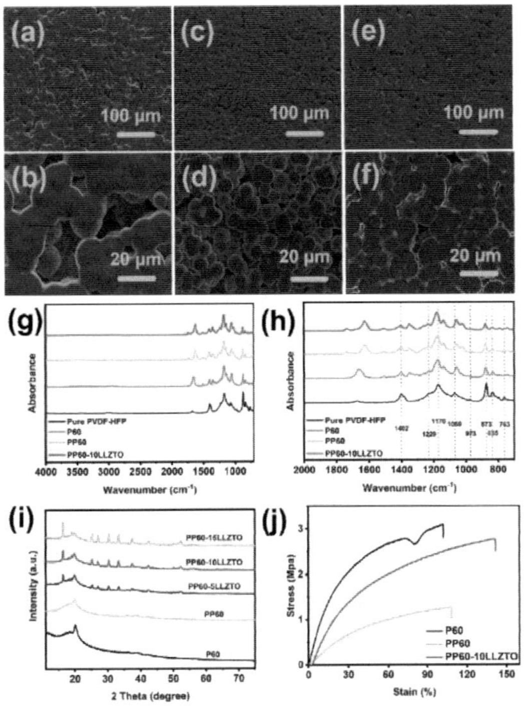

Figure 3. (a, b) SEM images of (a, d) P60, (c, d) PP60, and (e, f) PP60-10LLZTO electrolytes; FTIR spectra of Pure PVDF-HFP, P60, PP60, and PP60-10LLZTO electrolytes in the ranges of (g) 4000~700 cm^{-1} and (h) 2000~700 cm^{-1}, (i) XRD plots of P60, PP60, PP60-5LLZTO, PP60-10LLZTO, and PP60-15LLZTO electrolytes, (j) tensile stress-strain curves of P60, PP60, and PP60-10LLZTO electrolytes.

To investigate the impact of PVAc and LLZTO incorporation on the PVDF-HFP molecule, we characterized Pure PVDF-HFP, P60, PP60, and PP60-10LLZTO solid electrolytes using FTIR spectroscopy. Figure 3g displays the FTIR spectra in the range of 4000~700 cm^{-1}, while Figure 3h provides a local magnification of the FTIR spectra. The α-phase crystallization peaks of PVDF-HFP at 763 cm^{-1} and 1402 cm^{-1} are significantly reduced with the addition of LiTFSI, indicating a substantial reduction in the crystallinity of the PM. The further reduction in the characteristic peak at 1402 cm^{-1} upon LLZTO addition suggests complexation between the -OH on the surface of LLZTO and the -CF group of PVDF-HFP. This disturbance in the ordered molecular arrangement of PVDF-HFP further decreases the crystallinity of the PM. The characteristic peaks at 1069 and 1170 cm^{-1}, corresponding to the symmetric stretching of -CF2, exhibit varying degrees of shifting after the addition of PVAc and further shifting when LLZTO is introduced. This shift indicates interactions between PVAc and LLZTO with PVDF-HFP. Furthermore, the characteristic peak of TFSI- at 1635 cm^{-1} undergoes significant shifting after the addition of PVAc, suggesting a robust interaction between PVAc and LiTFSI, promoting the dissolution and dissociation of LiTFSI.

To further verify the effects of PVAc and LLZTO on the crystallinity of the PVDF-HFP matrix, XRD diffraction characterization was conducted on P60, PP60, PP60-5LLZTO, PP60-10LLZTO, PP60-15LLZTO. Figure 3i illustrates that the characteristic peaks located at 18.4° and 38.9° significantly diminish when PVAc is added, accompanied by a substantial decrease in the primary PVDF-HFP peak at 20.2°. This reduction indicates that the addition of the amorphous polymer PVAc can disrupt the ordered arrangement of the PVDF-HFP molecular chains, resulting in a decrease in the crystallinity of the PM. With the further addition of LLZTO to the solid-state electrolyte, the main peak of the β-phase of PVDF-HFP at 20.2° decreases as the LLZTO content increases. This suggests that the synergistic effect of PVAc and LLZTO further expands the amorphous zone in the composite solid-state electrolyte, consistent with the FTIR results.

In solid-state lithium metal batteries, the mechanical strength of the solid-state electrolyte is particularly critical. A solid-state electrolyte with excessively high mechanical strength may struggle to establish close contact with the electrode material, resulting in a large interfacial impedance. Conversely, a solid-state electrolyte with insufficient mechanical strength may struggle to resist the puncture of lithium dendrites. Figure 3j displays the stress-strain curves of P60, PP60, and PP60-10LLZTO electrolyte membranes Notably, the P60 electrolyte membrane exhibits the highest tensile stress (3.09 MPa), while the addition of PVAc to the PP60 electrolyte membrane significantly reduces its overall mechanical strength to 1.28 MPa. However, the elongation at break slightly increases compared to the P60 electrolyte film, indicating that the addition of PVAc enhances the flexibility of the electrolyte film, promoting close contact with the electrode material and reducing interfacial impedance. The introduction of LLZTO increases the tensile stress of the PP60-10LLZTO electrolyte film to 2.76 MPa, though it remains lower than that of the P60 electrolyte film. Nonetheless, the PP60-10LLZTO electrolyte film exhibits the highest elongation at break, measuring 141.6%. This implies that the synergistic effect of LLZTO and PVAc significantly improves material flexibility, facilitating close contact between the electrolyte film and the electrode material. It can also buffer the volume expansion of lithium dendrites during charging and discharging, thereby improving the cycling performance of the battery.

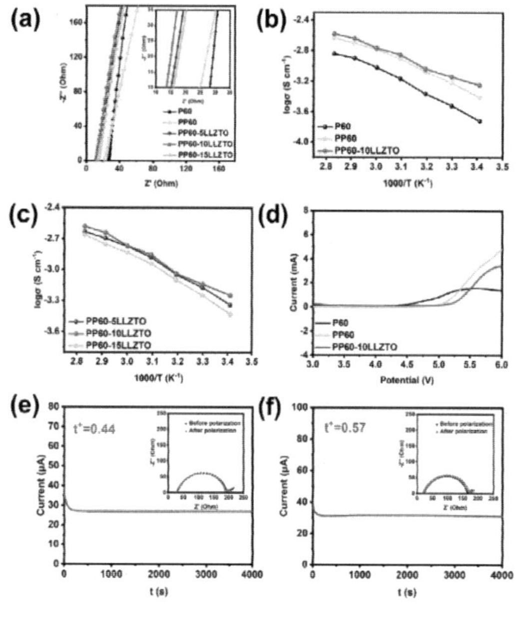

Figure 4. (a) Nyquist plots of P60, PP60, PP60-5LLZTO, PP60-10LLZTO, and PP60-15LLZTO electrolytes at 30 °C, (b) Arrhenius curves of P60, PP60, and PP60-10LLZTO electrolytes, (c) CSEs with different LLZTO contents of the Arrhenius curves, (d) LSV curves of P60, PP60, PP60-10LLZTO electrolytes, (e) chronoamperometric curves of PP60 and (f) PP60-10LLZTO electrolytes at 10 mV (insets show electrochemical impedance spectra before and after polarization).

The electrical resistance of different electrolytes can be measured by electrochemical impedance spectroscopy, and the IC of the electrolyte can be computed using Equation (1). Figure 4a shows the

EIS impedance spectra of different electrolytes at 30 °C. When PVAc is added to PVDF-HFP, there is a significant decrease in its impedance. This is due to the fact that PVDF-HFP can be mutually attracted to PVAc, thus disrupting the intermolecular ordering of PVDF-HFP, diminishing the crystallinity of the PM and increasing the amorphous region. Further incorporation of LLZTO into the copolymer solid electrolyte results in a continued decrease in impedance, with PP60-10LLZTO exhibiting the highest IC at $7.23*10^{-4}$ S/cm. This may be attributed to the fact that the incorporation of cubic-phase LLZTO provides more ionic transport channels for Li^+, further reducing the crystallinity of the solid electrolyte and increasing the ion transport rate.

The Arrhenius curves of P60, PP60, PP60-5LLZTO, PP60-10LLZTO, and PP60-15LLZTO were constructed by adjusting the temperatures (Figures 4b and 4c), and the activation energies of different electrolytes were determined by fitting them according to the Arrhenius equation (Eq. 2) (Table 1). Among them, PP60-10LLZTO exhibits the lowest E_a value (0.105 eV), indicating its enhanced favorability for Li^+ transport and the operation of solid-state lithium metal batteries at room temperature. However, as the LLZTO content increases to 15 wt.%, the IC of PP60-15LLZTO begins to decline (Figure 4c). This decrease is attributed to the excessive agglomeration of LLZTO in the polymer matrix, hindering ion transport. Therefore, 10 wt.% of LLZTO was chosen as the optimal addition amount.

Table 1 Activation energies of different electrolytes

solid state electrolyte	E_a /eV
P60	0.135
PP60	0.117
PP60-5LLZTO	0.108
PP60-10LLZTO	0.105
PP60-15LLZTO	0.115

To examine the electrochemical stability window of different solid-state electrolytes, we assembled P60, PP60, and PP60-10LLZTO solid state electrolytes with SS/solid state electrolyte/Li batteries. The evaluation of these solid-state electrolytes was conducted using linear voltammetric scanning (LSV) at 30 °C, as depicted in Figure 4d. The oxidation of the P60 electrolyte initiates around 4.6 V, while the PP60 copolymer solid-state electrolyte, with the addition of PVAc, begins oxidation at approximately 5.0 V. Notably, the oxidation potential of the PP60-10LLZTO composite solid-state electrolyte, with additional LLZTO, is around 5.25 V. The superior electrochemical stability exhibited by PP60-10LLZTO may be attributed to the synergistic presence of PVAc and LLZTO, enhancing the antioxidant capacity of the PVDF-HFP matrix and expanding its electrochemical stability window. This characteristic is advantageous for the application of the electrolyte in high-voltage cathode materials.

In Figures 4e and 4f, chronoamperometric curves and impedance comparisons before and after polarization are shown for PP60 and PP60-10LLZTO solid-state electrolytes, measured at an applied voltage of 10 mV (inset). Calculations based on Eq. (3) indicate that the Li^+ transference number of PP60 at 30 °C is approximately 0.44. This is mainly due to the fact that the addition of the amorphous polymer PVAc can disrupt the ordered arrangement of the molecular chain of PVDF-HFP and reduce the crystallinity, thus enhancing the migration rate of Li^+. In contrast, the Li^+ transference number of the PP60-10LLZTO composite solid-state electrolyte, with further LLZTO addition, is as high as 0.57, surpassing that of the PP60 electrolyte (0.44). The main reasons for this are as follows: firstly, the introduction of LLZTO promotes the dissociation of lithium salts, providing more Li^+, and further expanding the amorphous zone of the PM; secondly, LLZTO, as an active filler, provides additional

Li^+ transport channels; finally, the abundant Lewis acid sites on the surface of LLZTO attract the $TFSI^-$, hindering its free movement and promoting the increase in the Li^+ transference number.

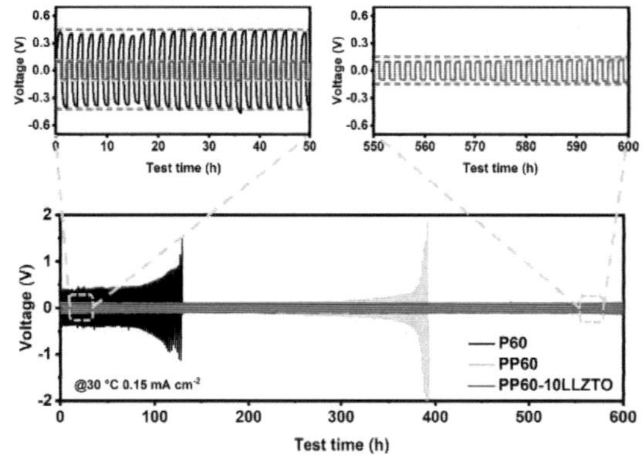

Figure 5. Constant current charging and discharging curves of Li/P60/Li, Li/PP60/Li and Li/PP60-10LLZTO/Li batteries at 30 °C 0.15 mA/cm^2.

Figure 5 displays the constant current charging and discharging curves of Li/Li symmetric cells assembled with P60, PP60 and PP60-10LLZTO electrolytes at 30 °C and 0.15 mA/cm^2 current density. In the figure, the initial polarisation potential of the Li/P60/Li cell is approximately 0.4 V, exhibiting higher polarisation compared to the Li/Li symmetric cell assembled with PP60 and PP60-10LLZTO electrolytes. The polarization potential of the Li/P60/Li cell experiences a significant increase after about 90 hours of cyclin. This is attributed to micro-short circuits caused by the uneven deposition of lithium ions due to the lower IC of the P60 electrolyte. This is attributed to the micro-short circuit caused by the uneven deposition of lithium ions due to the lower IC of the P60 electrolyte. The Li/PP60/Li cell, incorporating PVAc, reduces the crystallinity of the PVDF-HFP matrix, enhancing the flexibility of the electrolyte. This results in a tight bond with lithium metal and a reduced polarisation potential of about 0.1 V. However, a notable increase in overpotential occurs after 300 hours of stable cycling. In contrast, the Li/Li symmetric battery composed of PP60-10LLZTO composite solid electrolyte, with both PVAc and LLZTO added, exhibits stable cycling for over 600 hours under the same conditions, with a polarisation potential consistently at about 0.11 V. The enhanced stability is attributed to the PP60-10LLZTO composite solid electrolyte's high mechanical strength, resisting lithium dendrites, and flexibility, providing better interfacial compatibility with lithium metal. Furthermore, the polarisation potential of Li/PP60-10LLZTO/Li cells during lithium deposition/stripping slightly decreases at the cycle's start. This phenomenon indicates the formation of a stable SEI film at the electrolyte membrane's interface with lithium metal during the initial lithium deposition/stripping process, demonstrating the good compatibility of PP60-10LLZTO with lithium metal and its effective contribution to the cycling stability of all-solid-state lithium-metal batteries.

To investigate the impact of the solid-state electrolyte on the cycling performance of LFP/Li batteries, we assembled LFP/PP60/Li and LFP/PP60-10LLZTO/Li batteries, respectively, and tested their cycling performance at a constant current of 0.5 C at 30 °C. In Figure 6a, the cycling performance of the LFP/PP60/Li battery reveals an initial discharge specific capacity of 120 mAh/g after three cycles of low-current activation. This capacity gradually increases in subsequent cycles, primarily due to the incomplete activation of the electrode materials at the 30 °C temperature. Over the cycling period, activation becomes progressively and fully effective, resulting in a capacity retention of 74.4% after 450 cycles. Figure 6c illustrates the LFP/PP60-10LLZTO/Li battery, showing that after three cycles of activation, the initial discharge specific capacity reaches an impressive 139 mAh/g. Even after 450 cycles, the discharge specific capacity remains at 124.3 mAh/g, with a capacity retention rate of 89.4% and a high Coulombic efficiency of 100%. Figures 6b and d present the voltage-capacity plots of the

LFP/PP60/Li and LFP/PP60-10LLZTO/Li batteries under different cycle times, respectively. The graphs reveal slight increases in the charging and discharging voltage plateaus during cycling, indicating that the addition of PVAc effectively reduces the polarization of the solid electrolyte. The outstanding cycling performance of the LFP/PP60-10LLZTO/Li batteries can be attributed to several factors. Firstly, the introduction of the amorphous polymer PVAc interacts with the PVDF-HFP matrix, disrupting the ordered arrangement of polymer molecular chains, reducing crystallinity, and increasing the amorphous region, facilitating rapid ion transport. Secondly, PVAc enhances the flexibility of the PM, promoting favorable interfacial contact with the electrode material. Finally, the introduction of the active filler LLZTO further reduces the crystallinity of the PM, providing more Li^+ transport channels, increasing ionic conductivity, supporting uniform lithium ion deposition, and enhancing the cycling performance of all-solid-state lithium metal batteries.

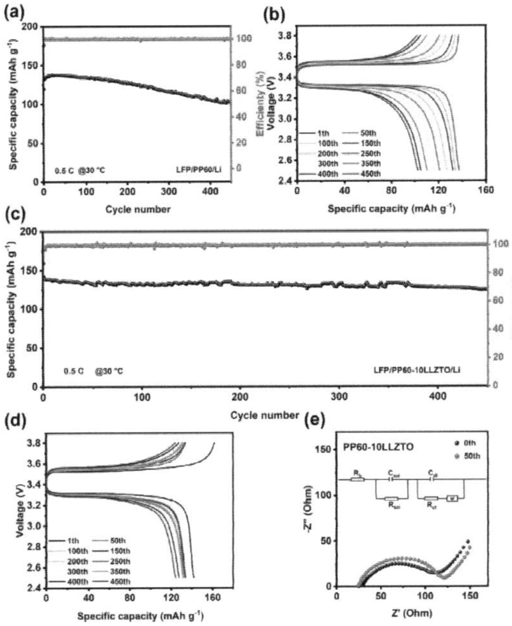

Figure 6. (a) Cycling performance (b) Voltage-capacity plot at different numbers of cycles of the LFP/PP60/Li battery, (c) Cycling performance (d) Voltage-capacity plot at different numbers of cycles of the LFP/PP60-10LLZTO/Li battery (e) Nyquist plot pre- and post-50 cycles.

Figure 6e presents the impedance comparison between the LFP/PP60-10LLZTO/Li battery before cycling at 30 °C and 0.5 C and after cycling for 50 cycles. The interfacial impedance R_{sei} of the LFP/PP60-10LLZTO/Li battery before cycling is only about 90 Ω, significantly lower than in similar works. The introduction of PVAc, which enhances the flexibility and viscosity of PVDF-HFP, plays a crucial role in better adherence to the electrode material, resulting in a substantial reduction in interfacial impedance. Although R_{sei} increased slightly after cycling, it indicates that during the cycling process, the composite solid electrolyte reacted with the electrode material at the interface, generating the SEI film.

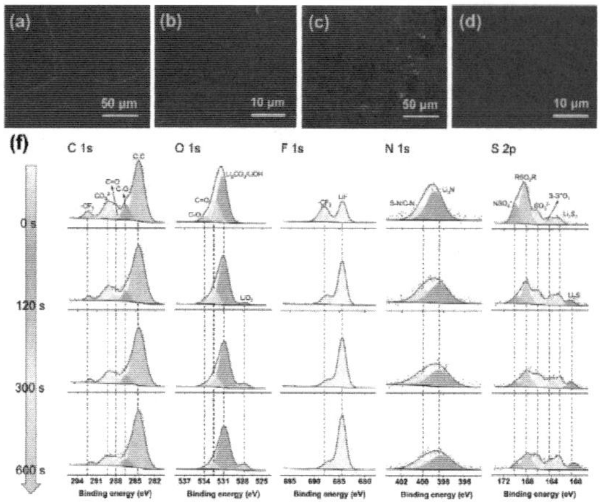

Figure 7 (a, b) SEM images of the uncycled lithium metal surface and (c, d) lithium metal anode surface of the LFP/PP60-10LLZTO/Li cell after 100 cycles, and (f) XPS spectra of the lithium anode (LA) surface of the LFP/PP60-10LLZTO/Li cell after 100 cycles at 30 °C and 0.5 C.

Figure 7 illustrates SEM images of the uncirculated lithium metal versus the lithium anode after 100 cycles. Figures 7a and 7b clearly depict the smooth and pristine surface of the uncirculated LA. In Figures 7c and d, a relatively flat SEI film is observed on the surface of the lithium metal anode after 100 cycles of the LFP/PP60-10LLZTO/Li cell, with no lithium dendrites or dead lithium formation. This suggests that the SEI film effectively inhibits lithium dendrite growth, reduces lithium loss, and enhances the cycling performance of the all-solid-state lithium metal battery. The interface between the solid-state electrolyte and the electrode material significantly influences battery performance. Figure 7f demonstrates the composition and chemical state of the SEI layer on the LA surface of the LFP/PP60-10LLZTO/Li battery after 100 cycles at 0.5 C and 30 °C with different etching times. In both C 1s and O 1s spectra, the intensities of the peaks (C-C, C=O, C-O, CO_3^{2-}, $-CF_3$, $LiOH/Li_2CO_3$) appear to be weakened to varying degrees after etching. After 120 s of etching, Li_2O peaks appear on the lithium metal surface, and their intensity increases with etching time. It is noteworthy that the presence of Li_2O favours lithium deposition. In F 1s and N 1s spectra, the intensities of LiF and Li_3N peaks increased and decreased with increasing etching time, respectively, indicating their role in the SEI film. In the S 2p spectrum, Li_2S is revealed only after a long period of etching, and its intensity continuously enhances with etching time. This suggests the formation of a double-layer SEI film on the lithium metal surface during cycling. The bottom layer, close to the LA, is mainly composed of inorganic components such as LiF, LiO_2, Li_2S, and Li_2S_2, promoting homogeneous lithium salt deposition and inhibiting lithium dendrite growth. The top layer, away from the LA, primarily consists of decomposition products of PVDF-HFP and PVAc. The formation of this double-layer SEI film greatly improves the stability of the electrode/electrolyte interface and enhances the cycling stability of all-solid-state lithium metal batteries.

4. Conclusion

In this study, a composite solid electrolyte, PP60-10LLZTO, was prepared by incorporating the amorphous polymer PVAc and the active filler LLZTO into the PVDF-HFP matrix. The addition of the amorphous polymer PVAc significantly reduced the crystallinity of the PVDF-HFP matrix, enhancing the flexibility of the PM. Building on this, the introduction of the active particle LLZTO further decreased the crystallinity of the PVDF-HFP matrix and impeded the movement of the large anion TFSI-. The resulting PP60-10LLZTO composite solid electrolyte exhibited excellent Li^+ transference number (0.57) and ionic conductivity ($7.23*10^{-4}$ S/cm at 30 °C). The LFP/PP60-

10LLZTO/Li battery demonstrated a discharge specific capacity of 139 mAh/g at 30 °C and 0.5 C during its initial cycle. Even after 450 cycles, it maintained a discharge specific capacity of 124.3 mAh/g, resulting in a capacity retention rate of 89.4%. Additionally, the battery exhibited a Coulombic efficiency of 100%. This study highlights the promising potential of incorporating PVAc and LLZTO into composite solid electrolytes for the development of advanced lithium metal batteries.

Notes

The authors declare no conflict of interest.

Acknowledgments

This work was financially supported by Key R&D Program of Zhejiang (Grant 2022C01033). The authors would like to express their gratitude to EditSprings (https://www.editsprings.cn) for the expert linguistic services provided.

Reference lists

[1] D.D. Han, Z.Y. Wang, G.L. Pan, X.P. Gao, Metal-Organic-Framework-Based Gel Polymer Electrolyte with Immobilized Anions To Stabilize a Lithium Anode for a Quasi-Solid-State Lithium-Sulfur Battery, ACS Appl. Mater. Interfaces 11(20) (2019) 18427-18435 https://doi.org/10.1021/acsami.9b03682

[2] C. Sun, J. Liu, Y. Gong, D.P. Wilkinson, J. Zhang, Recent advances in all-solid-state rechargeable lithium batteries, Nano Energy 33 (2017) 363-386 https://doi.org/10.1016/j.nanoen.2017.01.028

[3] L. Du, B. Zhang, W. Deng, Y. Cheng, L. Xu, L. Mai, Hierarchically Self-Assembled MOF Network Enables Continuous Ion Transport and High Mechanical Strength, Adv. Energy Mater. 12(24) (2022) 2200501-2200511 https://doi.org/10.1002/aenm.202200501

[4] A. Gupta, E. Kazyak, N.P. Dasgupta, J. Sakamoto, Electrochemical and Surface Chemistry Analysis of Lithium Lanthanum Zirconium Tantalum Oxide (LLZTO)/Liquid Electrolyte (LE) Interfaces, J. Power Sources 474 (2020) https://doi.org/10.1016/j.jpowsour.2020.228598

[5] M. Wu, D. Liu, D. Qu, Z. Xie, J. Li, J. Lei, H. Tang, 3D Coral-like LLZO/PVDF Composite Electrolytes with Enhanced Ionic Conductivity and Mechanical Flexibility for Solid-State Lithium Batteries, ACS Appl. Mater. Interfaces 12(47) (2020) 52652-52659 https://doi.org/10.1021/acsami.0c15004

[6] Q. Zhang et al., Chloride-Reinforced Solid Polymer Electrolyte for High-Performance Lithium Metal Batteries, ACS Appl. Mater. Interfaces 15(14) (2023) 18252-18261 https://doi.org/10.1021/acsami.2c20734

[7] H. Xie et al., Flexible, Scalable, and Highly Conductive Garnet-Polymer Solid Electrolyte Templated by Bacterial Cellulose, Adv. Energy Mater. 8(18) (2018) 1703474-1703481 https://doi.org/10.1002/aenm.201703474

[8] R. Fang, B. Xu, N.S. Grundish, Y. Xia, Y. Li, C. Lu, Y. Liu, N. Wu, J.B. Goodenough, Li(2) S(6) -Integrated PEO-Based Polymer Electrolytes for All-Solid-State Lithium-Metal Batteries, Angew Chem Int Ed Engl 60(32) (2021) 17701-17706 https://doi.org/10.1002/anie.202106039

[9] X. Wu, Y. Zheng, W. Li, Y. Liu, Y. Zhang, Y. Li, C. Li, Solid electrolytes reinforced by infinite coordination polymer nano-network for dendrite-free lithium metal batteries, Energy Storage Mater. 41 (2021) 436-447 https://doi.org/10.1016/j.ensm.2021.06.009

[10] W. Liu, C. Yi, L. Li, S. Liu, Q. Gui, D. Ba, Y. Li, D. Peng, J. Liu, Designing Polymer-in-Salt Electrolyte and Fully Infiltrated 3D Electrode for Integrated Solid-State Lithium Batteries, Angew Chem Int Ed Engl 60(23) (2021) 12931-12940 https://doi.org/10.1002/anie.202101537

[11] X. Guo, Z. Ju, X. Qian, Y. Liu, X. Xu, G. Yu, A Stable Solid Polymer Electrolyte for Lithium Metal Battery with Electronically Conductive Fillers, Angew Chem Int Ed Engl 62(7) (2023) e202217538 https://doi.org/10.1002/anie.202217538

[12] T. Wang, X. Zhang, N. Yuan, C. Sun, Molecular design of a metal–organic framework material rich in fluorine as an interface layer for high-performance solid-state Li metal batteries, Chem. Eng. J. 451 (2023) https://doi.org/10.1016/j.cej.2022.138819

[13] Z. Wang, Q. Guo, R. Jiang, S. Deng, J. Ma, P. Cui, X. Yao, Porous poly(vinylidene fluoride) supported three-dimensional poly(ethylene glycol) thin solid polymer electrolyte for flexible high temperature all-solid-state lithium metal batteries, Chem. Eng. J. 435 (2022) 135106-135115 https://doi.org/10.1016/j.cej.2022.135106
[14] S. Wen et al. Deng, Integrated design of ultrathin crosslinked network polymer electrolytes for flexible and stable all-solid-state lithium batteries, Energy Storage Mater. 47 (2022) 453-461 https://doi.org/10.1016/j.ensm.2022.02.035
[15] K. Fu et al., Wachsman, L. Hu, Three-dimensional bilayer garnet solid electrolyte based high energy density lithium metal–sulfur batteries, Energy Environ. Sci. 10(7) (2017) 1568-1575 https://doi.org/10.1039/c7ee01004d
[16] R.-A. Tong, L. Chen, G. Shao, H. Wang, C.-A. Wang, An integrated solvent-free modification and composite process of $Li_{6.4}La_3Zr_{1.4}Ta_{0.6}O_{12}$/Poly(ethylene oxide) solid electrolytes: Enhanced compatibility and cycle performance, J. Power Sources 492 (2021) 229672-229681 https://doi.org/10.1016/j.jpowsour.2021.229672
[17] X. Zheng, T. Yang, J. Wei, C. Wang, M. Chen, Co-contribution of quenching and nanocrystallization on ionic-conductivity improvement of a composite electrolyte of polyethylene Oxide/$Li_7La_3Zr_2O_{12}$ nanofibers at 45 °C for all-solid-state Li metal batteries, J. Power Sources 496 (2021) 229843-229853 https://doi.org/10.1016/j.jpowsour.2021.229843
[18] S.-E. Lee, H.-T. Sim, Y.-J. Lee, S.-B. Hong, K.Y. Chung, H.-G. Jung, D.-W. Kim, Li_6PS_5Cl-based composite electrolyte reinforced with high-strength polyester fibers for all-solid-state lithium batteries, J. Power Sources 542 (2022) 231777-231786 https://doi.org/10.1016/j.jpowsour.2022.231777
[19] L. Wei, X. Xu, S. Jiang, K. Xi, L. Zhang, Y. Lan, J. Yin, H. Wu, Y. Gao, Zeolitic imidazolate framework upgrading polyethylene oxide composite electrolyte for high-energy solid-state lithium batteries, J. Colloid Interface Sci. 630 (2023) 232-241 https://doi.org/10.1016/j.jcis.2022.09.142
[20] E. Yi, H. Shen, S. Heywood, J. Alvarado, D.Y. Parkinson, G. Chen, S.W. Sofie, M.M. Doeff, All-Solid-State Batteries Using Rationally Designed Garnet Electrolyte Frameworks, ACS Appl. Energy Mater. 3(1) (2020) 170-175 https://doi.org/10.1021/acsaem.9b02101
[21] Q. Zeng et al., Cross-Linked Chains of Metal–Organic Framework Afford Continuous Ion Transport in Solid Batteries, ACS Energy Lett. 6(7) (2021) 2434-2441 https://doi.org/10.1021/acsenergylett.1c00583

AUTHOR INDEX